Essays on Infrastructure-as-code

Ravi Rajamani

Copyright © 2025 Ravi Rajamani.

All rights reserved. No part of this book may be used or reproduced by any means, graphic, electronic, or mechanical, including photocopying, recording, taping or by any information storage retrieval system without the written permission of the author except in the case of brief quotations embodied in critical articles and reviews.

Archway Publishing books may be ordered through booksellers or by contacting:

Archway Publishing
1663 Liberty Drive
Bloomington, IN 47403
www.archwaypublishing.com
844-669-3957

Because of the dynamic nature of the Internet, any web addresses or links contained in this book may have changed since publication and may no longer be valid. The views expressed in this work are solely those of the author and do not necessarily reflect the views of the publisher, and the publisher hereby disclaims any responsibility for them.

Any people depicted in stock imagery provided by Getty Images are models, and such images are being used for illustrative purposes only. Certain stock imagery © Getty Images.

ISBN: 978-1-6657-6846-7 (sc)
ISBN: 978-1-6657-6848-1 (hc)
ISBN: 978-1-6657-6847-4 (e)

Library of Congress Control Number: 2024924580

Print information available on the last page.

Archway Publishing rev. date: 01/30/2025

This book is dedicated to my teacher.

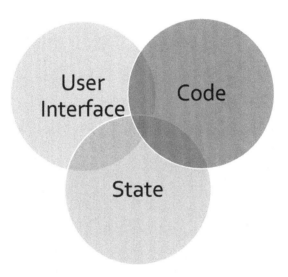

Figure shows the relationship between three representations of resources that must be kept in sync.

Contents

IaC Architecture 1
IaC Types 4
IaC Comparisons 7
IaC Automations 10
IaC Naming 12
IaC Characteristics 14
Resource Locking 16
Deadlocks 18
IaC Dependencies 20
IaC Complexity 23
Knowledge graphs 28
The Hidden Factor 30
IaC caveats 33
IaC Challenges 35
IaC Resolutions 37
IaC customizations 42
IaC for ML Pipelines 44
The case for an ML Pipeline 46
Workflows 53
Comparisons to Deis Workflow 57
Solution Accelerator 60
IaC Modernization 62
The refactoring of old code to new microservices 66
Maintainability, performance, and security 69
Path towards Microservices architecture 71
Application Migration Process 73
Incremental code migration strategy 75
Towards architecture driven modernization 77
Reverse engineering using models 80
Model-driven Software development 82
Rehost, Refactor, Rearchitect, Rebuild, Retire 87

A recommendation to businesses	89
Field Guide	93
Questionnaire	95
The Application Migration scenario for serving static content – a case study	102
Well-architected framework	104
The power of infrastructure consolidation	108
IaC Innovations	110
IaC regulation and compliance	114
Authorization	118
Privilege and permissions: A case study	121
Troubleshooting of role assignments via IaC: a case study	123
Automated Cloud IaC using Copilot	126
Emerging Trends and Generative AI	128
Popularity of chatbots	133
Managing copilots	139
LLM-as-a-judge	141
Data Infrastructure for Generative AI	148
Networking: a case study	152
Understanding Workloads for business continuity and disaster recovery (aka BCDR)	154
Workload 1: Applications	154
Workload #2: Hosting	156
Workload #3: Analytical workspaces:	157
Workload #4: Traffic	158
Workload #5: Data stores:	159
Ownership	161
A note about choices between public clouds	163
Just-in-time access	166
Technical Debt in IaC	169
DevOps for IaC	171
Top-Down vs Bottoms-up	173
Mitigation of political and regulatory risk in cloud infrastructure	175
Data Engineering	178
Support	189

IaC Architecture

A public cloud is a tiered implementation of proprietary investments in compute, storage, and networking from well-known providers such as Microsoft, comprising an Infrastructure-as-a-service aka IaaS layer, a Platform-as-a-service aka PaaS layer next, followed by resource manager and Dev-ops tools on top. The public cloud offers capabilities to the general public in the form of services from the provider's services portfolio that can be requested as instances called resources. It is for these resources that a certain form of manifests are recorded and saved as code and often referred to as Infrastructure-as-code or IaC for short. The capabilities of a public cloud are wide and deep targeting use cases in web and mobile, Internet of Things, Microservices, Data + analytics, Identity management, Media streaming, High Performance Compute and Cognitive computing. These services all utilize core investments in terms of compute, networking, storage and security, organized in a hierarchy of increasing scope in the form of datacenters at a physical location, usually the size of a football stadium, availability zones comprising one or more datacenters, a region such as West US or East US comprising one or more availability zones, and a global presence of individual resources spanning multiple regions. Both for the provider and the general public, IaC is a common paradigm for self-service templates to manage, capture and track changes to a resource during its lifecycle.

There is no denying that Infrastructure-as-code aka IaC can help to create and manage infrastructure with versioning, reusability, and sharing – all of which helps to provision resources quickly and consistently and manage them for their lifetime. But unlike software product code that is written for business purposes and provides a strong foundation for system architecture that weathers changing business requirements over time and often grows to become veritable platforms for ecosystems where independent vendors can join to increase the business value, IaC manifests in variations and combinations depending on environment, purpose, and scale. That said, as with all code, it encompasses complete development process and includes CI/CD platform, DevOps, and testing tools. The DevOps based approach is critical to rapid iteration and development cycles for improvements. This makes IaC spread over in a variety of forms. The more articulated the IaC the more predictable and cleaner the deployments.

The IaC architecture is almost always dominated by the choice of technology stacks. There is no universal system architecture or a single point of maintenance but a more DevOps oriented tailored approach with all the tools necessary to keep the deployments consistent and repeatable. Technology varies with cloud native forms, providers like Ansible, Terraform, and domain specific language such as Pulumi. IaC can be curated as a set of machine-readable files, descriptive models, and configuration templates. Then there are two approaches for writing it which are an imperative approach and a declarative approach. The imperative approach allows users to specify the exact steps to be taken for a change and the system does not deviate from them while a declarative approach specifies the final form, and the tool or platform involved goes through the motion of provisioning them.

Infrastructure can be made available as an online service and shared offline as code. Provisioning infrastructure can be a cloud service, and many public clouds offer it in their service portfolio. These so-called native infrastructures are great for leveraging the public cloud built-in features but more than usual, organizations build a veritable library of assets and prefer it not to be limited to any one cloud-based resources. It can even include on-premises infrastructure. No matter what choices are made and the decision process for navigating the IaC landscape, it is unquestionable that IaC reduces shadow IT within organizations, integrates directly with CI/CD platforms, version controlling infrastructure and configuration changes, standardizing infrastructure, effectively managing configuration drift and with the ability to scale up or out without increasing CapEx or OpEx.

Configuration Management is separate from infrastructure management although tools like Ansible provide hybrid solutions. True configuration management is demonstrated by software like CFEngine while infrastructure management is demonstrated by providers like Terraform and Pulumi. Businesses can mix and match any tool and use them in their CI/CD pipelines depending on their custom requirements.

As a real-world example, a developer writes application code and the configuration management related instructions that will trigger actions from the virtualization environment. When the code is delivered, configuration management and infrastructure management provide a live operational environment for testing. When the tests run and the error detection and resolution occur, the new code changes become ready for deployment to customer facing environments. Managing the state drift as changes keep propagating is one of the core management routines for Infrastructure-as-code.

IaC Types

There is a growing need for dynamic, dependable, and repeatable infrastructure as the scope of deployment expands from small footprint to cloud scale. Some of the manual approaches and management practices cannot keep up. There are two popular ways to meet these demands on the Azure public cloud which are provider-independent infrastructure-as-code aka IaC and public cloud specific resource templates. Both these approaches can provide deterministic and software-defined infrastructure, but they also have independent use cases. Specifically, there are use cases for DevSecOps and they apply to the development and operation of trustworthy infrastructure-as-a-code. The term DevSecOps stands for a framework that integrates security practices into every phase of the software development lifecycle.

One example of provider independent Infrastructure-as-code is Terraform which highlights the use case of universally extensible with provider specific templates that furnish IaC for resource types. It is a one-stop shop for any infrastructure, service, and application configuration. It can handle complex order-of-operations and composability of individual resources and encapsulated models. It is also backed by an open-source community for many providers and their modules with public documentation and examples. Public cloud providers like Microsoft also work directly with the Terraform maker on building and maintaining related providers and this partnership has gained widespread acceptance and usage. Perhaps, one of the best features is that it tracks the state of the real-world resources which makes Day-2 and onward operations easier and more powerful.

ARM templates are entirely from Microsoft that are consumed internally and externally as the de facto standard for describing resources on the Azure public cloud and come with their import and export options. This serves as a great example for public-cloud specific resource templates and throughout this book, these two types of IaC are used for explanations. There is a dedicated cloud service called the Azure Resource Manager service that expects and enforces this convention for all resources to provide effective validation, idempotency and repeatability.

Whenever there are templates, there are also Blueprints. Azure Blueprints, for example, can be leveraged to allow an engineer or architect to sketch a project's design parameters, define a repeatable set of resources that implements and adheres to an organization's standards, patterns, and requirements. It is a declarative way to orchestrate the deployment of various resource templates and other artifacts such as role assignments, policy assignments, ARM templates, and Resource Groups. Blueprint Objects are stored in the CosmosDB and replicated to multiple Azure regions. Since it is designed to setup the environment, it is different from resource provisioning. This package fits nicely into a CI/CD.

With Azure templates, one or more Azure resources can be described with a document, but it does not exist natively in Azure and must be stored locally or in source control. Once those resources are deployed, there is no active connection or relationship to the template.

Other IaC providers like Terraform also have features such that it tracks the state of the real-world resources which makes Day-2 and onward operations easier and more powerful and with Azure Blueprints, the relationship between what should be deployed and what was deployed is preserved. This connection supports improved tracking and auditing of deployments. It even works across several subscriptions with the same blueprint.

Typically, the choice is not between a blueprint and a resource template because one comprises the other but between an Azure Blueprint and a Terraform state. They differ in their organization methodology as top-down or bottom-up. Blueprints are great candidates for compliance and regulations while Terraform is preferred by developers for their flexibility.

Blueprints manage Azure resources only while Terraform can work with various resource providers.

Once the choice is made, some challenges will be tackled next. The account with which the IaC is deployed and the secrets it must know for those deployments to occur correctly are something that works centrally and not in the hands of individual end-users. Packaging and distributing solutions for end-users is easier when these can be read from a single source of truth in the cloud, so at least the location in the cloud for the solution to read and deploy the infrastructure must be known beforehand.

The DevSecOps workflow has a double loop between various stages including create->plan->monitor->configure->Release->Package->Verify where the create, plan, verify and package stages belong to Dev or design time and the monitor, configure and release belong to operations runtime. SecOps sits at the cusp between these two halves of Dev and Ops and participates in the planning, package, and release stages.

Some of the greatest challenges of DevSecOps are firstly, cultural in that it comes from market fragmentation in terms of IaC providers and secondly, variety of wide skills required for such IaC. Others include definition of well-known code or design patterns, difficulty in replicating errors, IaC language specifics and diverse toolset, security and trustworthiness, configuration drift and changing infrastructure requirements.

IaC Comparisons

When traditionally manual approaches to creation and management of infrastructure for organizational software assets, does not scale, a way to define Infrastructure-as-Code aka IaC for short, is necessary to meet the automation demands in favor of dynamic, dependable, and repeatable infrastructure. Among the popular choices for Azure public cloud are Terraform and ARM templates and Biceps, each of which have their own language and format and are distinct from one another.

The challenges and benefits of these are discussed now. Their main differences are called out first.

Terraform is universally extendable through providers that furnish IaC for resource types. It is a one-stop shop for any infrastructure, service, and application configuration. It can handle complex order-of-operations and composability of individual resources and encapsulated models. It is also backed by an open-source community for many providers and their modules with public documentation and examples. Microsoft also works directly with the Terraform maker on building and maintaining related providers and this partnership has gained widespread acceptance and usage. Perhaps, one of the best features is that it tracks the state of the real-world resources which makes Day-2 and onward operations easier and more powerful.

ARM templates is entirely from Microsoft consumed internally and externally as the de facto standard for describing resources on Azure and with their import and export options. There is a dedicated cloud service called the Azure Resource Manager service that expects and enforces this convention for all resources to provide effective validation, idempotency and repeatability.

In principle, they may both appear similar in their purpose to describe Infrastructure-as-Code but Microsoft owning this convention means that the public cloud will not do away with this format any time soon even as features are developed in newer formats such as Bicep. Bicep provides more concise syntax and improved type safety, but they compile to ARM templates which is the de facto standard to declare and use Azure resources and supported by the unified Azure Resource Manager. Bicep is a new domain-specific language that was recently developed for authoring ARM templates by using easier syntax. Bicep is typically used for resource deployments to Azure. It is a new deployment-specific language that was recently developed. Either or both JSON and Bicep can be used to author ARM templates and while JSON is ubiquitous, Bicep can only be used with Resource Manager Templates. In fact, Bicep has tooling that converts Bicep templates into standard Json Templates for ARM Resources by a process called transpilation. This conversion happens automatically but it can also be manually invoked. Bicep is succinct so it provides a further incentive. The use of built-in functions, conditions and loops for repetitive resources infuses logic into the ARM templates.

Other infrastructure providers like Kubernetes have a language that articulates state so that its control loop can reconcile these resources. The resources can be generated and infused with specific configuration and secret using a ConfigMap generator and a secret generator, respectively. For example, it can take an existing application.properties file and generate a ConfigMap that can be applied to new resources. Kustomization allows us to override the registry for all images used in the containers for an application. There are two advantages to using it. First, it allows us to configure the individual components of the application without requiring changes in them. Second, it allows us to combine components from different sources and overlay them or even override certain configurations. The kustomize tool provides this feature. Kustomize can add ConfigMaps and secrets to the deployments using their specific generators, respectively. Kustomize is static declaration.

It allows adding labels across components. We can choose the groups of Kubernetes resources dynamically using selectors, but they must be declared as yaml. This kustomization yaml is usually stored as manifests and applied on existing components so they refer to other yamls. Arguably, yaml is the most succint format of templates.

Azure Blueprints can be leveraged to allow an engineer or architect to sketch a project's design parameters, define a repeatable set of resources that implements and adheres to an organization's standards, patterns, and requirements. It is a declarative way to orchestrate the deployment of various resource templates and other artifacts such as role assignments, policy assignments, ARM templates, and Resource Groups. Blueprint Objects are stored in the CosmosDB and replicated to multiple Azure regions. Since it is designed to setup the environment, it is different from resource provisioning. This package fits nicely into a CI/CD pipeline and handles both what should be deployed and the assignment of what was deployed.

IaC Automations

Automations pave the way for zero-touch, robust and fully managed deployments, integrations and customized problem solving. For example, a dedicated cloud service cam deploy other cloud services and their associated resources as well as the infrastructure for customers who require deployment stamps comprising of those resources. Such a service that deploys other services or resources, must accept IaC deployment logic with templates, intrinsics, and deterministic execution that works much like any other workflow management system. This helps to determine the order in which to run them and with retries. The tasks are self-described. The automation consists of a scheduler to trigger scheduled workflows and to submit tasks to the executor to run, an executor to run the tasks, a web server for a management interface, a folder for the directed acyclic graph representing the deployment logic artifacts, and a metadata database to store state. The workflows don't restrict what can be specified as a task which can be a built-in or custom Operator, a Sensor which is entirely about waiting for an external event to happen, and a Custom task that can be specified via a Python function decorated with a @task.

The organization of such artifacts pose two necessities. First, certain definitions, phases and sequences need to be declared to leverage the built-in templates and deployment capabilities of the target IaC provider as well as their packaging in the format suitable to the automation. Second the co—ordination of context management switches between automation service and IaC provider. This involved a preamble and an epilogue to a context switch for bookkeeping and state reconciliation.

This taught us that large IaC authors are best served by uniform, consistent and global naming conventions, registries that can be published by the system for cross subscription and cross region lookups, parametrizing diligently at every scope including hierarchies, leveraging dependency declarations, and reducing the need for scriptability in favor of system and user defined organizational units of templates. Leveraging supportability via read-only stores and frequently publishing continuous and up-to-date information on the rollout helps alleviate the operations from the design and development of IaC.

IaC writers frequently find themselves in positions where the separation between pipeline automation and IaC declarations are not clean, self-contained or require extensive customizations. One of the approaches that worked on this front is to have multiple passes on the development. With one pass providing initial deployment capability and another pass consolidating and providing best practice via refactoring and reusability. Enabling the development pass to be DevOps based, feature centric and agile helps converge to a working solution with learnings that can be carried from iteration to iteration. The refactoring pass is more generational in nature. It provides cross-cutting perspectives and non-functional guarantees.

A library of routines, operators, data types, global parameters and registries are almost inevitable with large scale IaC deployments but unlike the support for programming language-based packages, these are organically curated in most cases and often self-maintained. Leveraging tracking and versioning support of source control, it is possible to provide compatibility as capabilities are made native to the IaC provider or automation service.

IaC Naming

If you know of parents who named their children as Child-surname-boy-01 and Child-surname-girl-02, you know at least one of the parents is a public cloud infrastructure engineer. This form of naming convention is not limited to one public cloud and is even recommended by IaC providers who provide a unifying language over multiple clouds.

Organizing the resources when they number in tens or more, demands a certain discipline as it would in library science. The public clouds suggest a naming convention that speaks to resource type, environment, workload, region, and instance so that there is no need to make an additional call to get the details of the resource to tell them apart. Unique names are also necessary when resources get destroyed and re-created many times. When we freeze the instance, customers can expect to find the resource the way they saw it earlier. Consider a database server that was provisioned and used to save some data, but it got destroyed and recreated and now the data does not exist. Customers referring to that server by its name might find data at one point and find it missing later. To ensure that the instances have changed, it is better to increment the instance number whenever a resource is destroyed and recreated.

It is also important to know that the naming conventions help to sort and index just as any data in a database. These are two powerful techniques when we want to retrieve a subset of resources among hundreds and find it easier to discern them. Closely related names are especially helpful to indicate a similarity in origin or workload and being able to spot similarity and differences are probably one of the core duties of building out infrastructure in an organization.

Another habit of infrastructure engineers deserves special attention. When trying to use a name as a literal to input on an editor or search a set of resources, the veterans will never try to interpret the names to reconstitute it part-by-part at another location, because such steps are prone to mistakes and typos. Instead, they will treat the name as opaque between copying and pasting and taking care that the name was selected whole. Unfortunately, a significant source of errors during deployments of large-scale infrastructure occurs because a resource or its component or associations were not referred to by the right name. It would help professionals deploying infrastructure-as-code to pay a lot of attention to names throughout the code just like they would if they were reading quantitative information from a resume.

Conventions can be strict or relaxed and allow grading to enforce order and structure and the inclusivity is up to the discretion of engineers as some may choose to combine or omit some components. Clear, consistent, and descriptive names will make resource management easier.

Another popular principle is the Copy-on-write where two entities sharing the same list of names will make a copy whenever one of the tries to modify it so that the other can continue to read the original. This principle applies even to resources that are currently being used so that the original set of resources are not destroyed until the traffic moves to the newly created resource set. Since creation and deletion can be repeated in an idempotent manner, this principle facilitates a safe, reliable, and robust migration.

IaC Characteristics

One of the commonly encountered situations is when IaC must be defined differently for non-production and production environments. There is a separation that must be maintained between actions taken on non-production and production environments because they require different maintenance. The diligence and rigor for production environments is usually high.

Although software product code and infrastructure-as-code can both be written with reusable modules and a centralized repository and pipeline, Infrastructure can vary between business objectives, and between environments. This results in distinct sets of resources and as such have their own lifecycle and maintenance requirements. Therefore, while the emphasis with software development has been one of system architecture and microservices framework, that for infrastructure is about variations and independent management. This calls for separation not only in declaration and definition but also in pipelines and resource deployments.

It is important to call out that this requirement to keep different sets of infrastructure resources available for different purposes also manifests in IaC as separate folders for various business objectives. Each folder will have its own set of templates, variables, parameters and so on and constitute a logical holistic declaration of all the resources used towards that objective. This might result in an explosion of folders and organizational units within them which makes it difficult to enforce consistency and best practices. The increase in folders must also be matched with investments that enforce consistency possibly as pipeline automation if they fall outside what the compiler can support. Investments in pipeline automations and tests or validations are just as important as they are with the code for software products.

Some of the hierarchy is determined by the IaC compiler and all most favor locality with all the resources and associated definitions to be available within the same folder for generating a plan. The restriction comes from the compiler requiring a root folder for the project to build. In some cases, there are features that allow import by virtue of referencing external modules. A common theme in organizing IaC is the use of common modules that act like a wrapper over primitives so that all the consumers from various projects have dependencies on single point of definitions. This is great for enforcement of consistency as well as introduction of optional attributes to resources. Large IaC assets also manifest maturity in their naming conventions, terse definitions and avoidance of unnecessary declarations and dependencies. This suits IaC because the unit of declarations is usually on a resource-by-resource basis.

Another frequently borrowed functionality from automations is scriptability. While IaC manifests the resource declarations, scriptability is sometimes unavoidable when working with various resources, not all of which have feature parity with IaC syntax. This calls for scripts to be made part of IaC and the use of pseudo-resources for this purpose is even facilitated by the compiler. However, it is important to remember that the idempotent and deterministic nature of IaC wins over the changes that scripts go through.

Resource Locking

Locks, policies and IaC sometimes compete to provide protection to resources and can overstep each other with conflicts that require resolution. Each of them has a part to play and cannot be done away with and the hope is that each pipeline run is smooth and leaves a clean state after its run.

This might be wishful thinking when resources that need to be created, modified, or deleted have sub-resources or are associated with other resources. With such dependencies, operations might result in an error that states that one or the other has locks on it. Locking is essential to prevent any accidental modifications to the resources. It is assumed that authorized operations will be able to unlock the resources prior to the change and then lock afterwards. With the example of a private endpoint on an Azure public cloud resource to provide private IP address for incoming traffic, associated resources including the parent resource, private links and dns zones might all get locked. Only when all the locks are released will the operations succeed. This makes it hard to know upfront which locks to acquire and release.

One of the approaches with pipeline automations is the cascaded unlocking of all resources in a resource hierarchy such as a resource group or subscription level. Since the identity with which the Azure operations are performed must be privileged. Only the Owner and User Access Administrator built-in roles can create and delete management locks. The corresponding permissions belong to the Microsoft.Authorization/* or Microsoft.Authorization/locks/* organizational prefix. Custom roles having these permissions could also be sufficient. It might be time consuming to go through all the resources and sub-resources in a resource hierarchy to unlock them first before the operations begin and to lock them at the end and often includes some wait time to be specified in the script. But this leaves the resources in a clean state for the changes to be propagated from the IaC to the management portal for these resources. It is also possible to conditionally run these for changes that carry certain labels or distinguishing features such as a filter on operations.

A policy might act like a catch-all to apply locking where locks are missed out from resources but a policy on the Azure public cloud has a compliance interval of 24 hours. It is also a default allow and explicit deny system. If a resource violates a policy, it is marked as non-compliant. The effects that a policy takes are detection or prevention. The IaC code is the ultimate source of truth for the resources and there are ways to specify locks in the IaC for resources that must behave independently from the collective approach taken by the policy. Anytime a policy changes the locks and the IaC is unaware, there is a conflict. It is preferable to keep locking as simple as possible without any customizations for any subset of resources so that the pipeline automation is sufficient to co-ordinate the locking and unlocking.

Finally, it is much easier to do locking and unlocking with command-line interface than execute it elsewhere. Both pipeline scripts and public cloud automations can execute these commands and although Runbooks might not be able to execute them in PowerShell, the az cli can certainly be run via functions or such other resources. Invoking a script for locking or unlocking does not require a resource or its state to change.

Deadlocks

This section discusses the resolution for the case when changes to resources involves breaking a deadlock in state awareness between, say a pair.

Let us make a specific association between say a firewall and a network resource such as a gateway. The firewall must be associated with the gateway to prevent traffic flow through that appliance. When they remain associated, they remember the identifier and the state for each other. Initially, the firewall may remain in detection mode where it is merely passive. It becomes active in the prevention mode. When the modes are attempted to be toggled, the association prevents it. Neither end of the association can tell what state to be in without exchanging information and when they are deployed or updated in place, neither knows about nor informs the other.

There are two ways to overcome this limitation.

First, there is a direction established between the resources where the update to one forcibly updates the state of the other. This is supported by the gateway when it allows the information to be written through by the update in the state of one resource.

Second, the changes are made by the IaC provider first to one resource and then to the other so that the update to the other picks up the state of the first during its change. In this mode, the firewall can be activated after the gateway knows that there is such a firewall.

If the IaC tries to form an association while updating the state of one, the other might end up in an inconsistent state. One of the two resolutions above works to mitigate this.

This is easy when there is a one-to-one relationship between resources. Sometimes there are one-to-many relationships. For example, a gateway might have more than a dozen app services as its backend members and each member might be allowing public access. If the gateway must consolidate access to all the app services, then there are changes required on the gateway to route traffic to each app service as intended by the client and a restriction on the app services to allow only private access from the gateway.

Consider the sequence in which these changes must be made given that the final operational state of the gateway is only acceptable when all barring none remain reachable for a client through the gateway.

If the app services toggle the access from public to gateway sooner than the gateway becomes operational, there is some downtime to them, and the duration is not necessarily bounded if one app service fails to listen to the gateway. The correct sequence would involve first making the change in the gateway to set up proper routing and then restricting the app services to accept only the gateway. Finally, the gateway validates all the app service flows from a client before enabling them.

Each app service might have nuances about whether the gateway can reach it one way or another. Usually, if they are part of the same virtual network, then this is not a concern, otherwise peering might be required. Even if the peering is made available, routing by address or resolution by name or both might be required unless they are universally known on the world wide web. If the public access is disabled, then the private links must be established, and this might require both the gateway and the app service to do so. Lastly, with each change, an app service must maintain its inbound and outbound properly for bidirectional communication, so some vetting is required on the app service side independent of the gateway.

Putting this altogether via IaC requires that the changes be made in stages and each stage validated independently.

IaC Dependencies

Among the frequently encountered disconcerting challenges faced by engineers who deploy infrastructure is the way to understand, capture and use dependencies. Imagine a clone army where all entities look alike and a specific one or two need to be replaced. Without having a name or identifier at hand, it is difficult to locate those entities, but it becomes even harder when we do not know which of the others are actually using them, so that we are mindful of the consequences of replacements. Grounding this example with cloud resources in azure public cloud, we can take a set of resources with a private endpoint each that gives them a unique private IP address, and we want to replace the virtual network that is integrated with these resources. When we switch the virtual network, the old and the new do not interact with one another and traffic that was flowing to a resource on the old network is now disrupted when that resource moves to a different virtual network. Unless we have all the dependencies known about who is using the resource that is about to move, we cannot resolve the failures they might encounter. What adds to the challenge is that the virtual network is like a carpet on which the resources stand, and this resource type is always local to an availability zone or region so there is no built-in redundancy or replica available to ease the migration. One cannot just move the resource as if it were moving from one resource group to another, it must be untethered and tied to another virtual network with a delete of the old private endpoint and the addition of a new. Taking the example a little further, IaC does not capture dependencies between usages of resources. It only captures dependencies on creation or modification. For example, a workspace that users access to spin up compute and run their notebooks. might be using a container registry over the virtual network but its dependency does not get manifested because the registry does not maintain a list of addresses or networks to allow. The only way to reverse-engineer the listing of dependencies is to check the DNS zone records associated with the private endpoint and the entries added to the callers that resolve the container registry over the virtual network. These entries will have private IP addresses associated with the callers and by virtue of the address belong to an address space designated to a sub-network, it is possible to tell whether it came from a connection device associated with a compute belonging to the workspace. By painful enumeration of each of these links, it is possible to draw a list of all workspaces using the container registry. These records that helped us draw the list may have a lot of stale entries as the callers disappear but do not clean up the record. So, some pruning might be involved, and it might change over time, but it will still be handy.

Dependencies between instances of the same resource types go undetected in Infrastructure-as-code aka IaC but are still important to resource owners. The knowledge that two resources of the same resource type have a dependency as a caller-callee cannot remain tribal knowledge and impacts the sequence at the time of both creation and destruction. Different IaC providers have different syntax and semantics associated with expressing dependencies, but no one can do away with it. At the same time, their documentation suggests using these directives as a last resort and often for one resource type dependency on another. In such cases, some prudence is necessary.

When the dependency is an entire module, this directive affects the order in which the deployment rolls out. The IaC runtime will process all the resources and data sources associated with that module. If the resource requires information generated by another resource such as its assigned and dynamic public IP address, then those references can still be made part of the attributes of this resource without requiring the directive to declare dependencies. The runtime will know that the references imply and implicit dependency. In such cases, it is not necessary to manually define the dependencies between other resources. When the dependencies are hidden such as access control policies must be managed and actions must be taken that require those policies to be present, then the directive becomes necessary. This directive does not impact replacements when the dependency undergoes a change. For that reason, a different directive is specified to cascade parent resource replacements when there is a change to the referenced resource or attribute.

At the time of deployment, none of the resources are operational. So, caller-callee relationships do not justify a depends_on directive to be specified. Also, if for any reason one resource was required to be present for the other to be created, the idempotency of the IaC allows the deployment to be run again so that if that creation order is not met, it will succeed the next time over because one of the two resources would go through since there is at least one that does not have a dependency. If a dependency must still be specified to get the order right the first time between resources of the same resource type, it is possible to sequentially specify them in the IaC source code. Finally, if the program order is not maintained correctly, it should be possible to introduce pseudo-attributes to these two resources of the same resource type that define different references to other hybrid resource types that have predetermined order established by virtue of being different resource types. These marginal references can be made to local-only resources such as those for generating private-keys, issuing self-signed TLS certificates, and even generating random ids. These serve as the same glue to help connect "real" infrastructure objects. These local resources have the added advantage of being shallow as in being visible only to the IaC runtime and not the cloud as well as being persisted only in the state referenced by the runtime.

Finally, the dependency information can indeed be used as a last resort by storing all dependencies in a separate store that can be queried dynamically at the time that the dependency information becomes relevant.

IaC Complexity

Software quality is often analyzed by static tools as early as possible in the development and deployment process. For example, one metric named cyclomatic complexity is a source code complexity metric that could indicate potential coding errors. It is calculated from linearly independent paths through a program module. It can be measured by static analysis tools like NDepend for .Net code that offers this and more analysis ranging from dependency visualization to quality gates and smart technical debt estimation. Unfortunately, IaC does not have similar analysis and insights. However, a new metric could help to determine deployment code organization errors from dependency graphs between resources, modules, and higher layer units of organization. It can be called IaC Complexity and it attempts to eliminate the drawbacks from well-known mistakes such as overly reuse of a single component that is prone to single point of failures – a trap that escapes those practicing Don't-repeat-yourself thumb rules.

Functions and Components do not require stack traces, exceptions, and error messages to determine the problematic code dependencies from runtime. They can be done as early as compile time, which forms the basis for introducing static dependency analysis of IaC code. This IaC complexity metric is merely a quantitative scoring of the dependencies so that the order of deployment is more prioritized leveraging the best practice of repeatability and idempotency. It does not suggest, for instance, to have multiple identity-access-management modules on the control path. It attempts to provide visualizations of all the usages so that components can be re-organized. Some can be moved internally, some can be avoided or some that that are internal can be moved outside the trust boundary. Components could be grouped behind a common interface to let different implementations be associated or called. Each usage of such a variation in components could be called out via tagging. In that sense, it allows developers to find and identify units of organization where the diligence paid to http proxies can be applied to the usages of this potentially hot-spot module.

Assuming linear relationship between units of organization, the metric can be calculated similar to page ranking. The only difference between well-known Google's algorithm and its application here is that this is not about popularty of web-pages as is an indicator of potential problem and hence the reference to the final score as a cost rather than ranking. Costing of the usages is a technique to utilize the structure of inbound and outbound references to calculate the cost recursively. A given unit is dependent on a set of units and each of those origins contributes to it. If the origin has N outbound links, it assigns a score of 1/N for the dependency of the unit and zero otherwise. The scaling factor for a dependency of a unit is inversely dependent on the total number of outbound edges of the origin for that dependency. This allows the rank or cost of the given unit to be computed as a simple sum of all the scaled costs of the units on which it is dependent. The sum must be adjusted by a constant factor to accommodate the fact that there are units that may have no forward dependencies.

This can be written in the form of linear equations as

$M * PR = (1-d)$

where $0 < d < 1$ and this denotes a damping factor,

PR is a N-dimensional vector and M is a N x N matrix. N is the number of units within the system.

The I'th component of the vector PR is PR_i and this is the pagerank of the unit i.

$M = 1 - d\, T$ where T is the transition matrix.

The components of T are given by the number of outgoing dependencies:

Tij = 1 / Cj (if unit j is dependent on unit i)

Tij = 0 (otherwise)

Cj is the number of dependencies on unit j.

The solution of the linear system of equations is

PR = (M-inverse) * (1-d)

The calculation of the inverse matrix M-inverse can be done numerically. Jacobii iteration is simple and suits this calculation well because it has exponential convergence.

PR{k+1} = (1 –d) + d T * PR{k} is the Jacobi iteration for k+1

PR{k} denotes the value of the PageRank vector after the k-th iteration. PR{0} = 1 is the initial PageRank vector.

Taking the result of the last calculation as the input for the new iteration reduces the number of iterations. The Jacobi iteration can also be written without the notation for iterations as

$PR_{xi} = (1 - d)/N + d(\frac{PR_{x1}}{C_{x1}} + \cdots + \frac{PR_{xn}}{C_{xn}})$ in its well-recognized form

Assuming non-linear complexity, we need something more than known inbound references and one that leverages latent contributions from the neighborhood with a set of parameters. The notion of cost or rank is not absolute but depends on the specific needs of the system or its usage. For example, the cost of migrating a unit given its dependencies can be computed with ranking algorithms like page rank as a relative cost to its neighborhood, but it might not even be significant for a system focused on a specific unit or subgraph. Since rank can be specialized by topic, user or query, we need something that can be extended across those boundaries. This is where vectors, neural net, embeddings and machine learning come useful. In fact, a class of ranking algorithms leveraging a neural network model can directly work on graphs and are called Graph Neural Networks. They take as input a graph G with a node for each unit and corresponding edges and a specific node n and return a score generalizing all forms of dependencies. Neural networks can unify topics and approaches. It is also a good general-purpose technique for ranking. As with all neural networks, they can be trained with samples including those where constraints such as the cost of a node i must be greater than the cost of node j, can be enforced. Vector v_n is assigned to a node n belonging to a vector space R and given the name of state for that node. A state is a collective representation of the node denoted by n from its neighborhood of nodes ne[n], a subset of the global set of nodes N.

$$x_n = \sum_{u \in ne[n]} h_w\, v_n,\, x_u,\, v_u,\quad n \in N$$

where h_w expresses the dependence of a node on its neighborhood and is parameterized by a set of features w. It is a generic non-linear function of the dependence between x_n and the states of the neighbors of n. This is sometimes called a feedforward neural network or a transition network or a hidden matrix.

The state x_n is the solution of the following system of equations:

1. A dependence function and
2. A learning function

The dependence function has an output o_n for each node n and belonging to a vector space R which depends on the state x_n and vector v_n. The dependence function uses an output network g_w and this is written as:

$$o_n = g_w(x_n, v_n),\quad n \in N$$

The above two equations define a method to produce an output rank o_n for each node given a graph.

The learning function to learn the parameters h_w and g_w is one that minimizes the error, and this error function can be some variation of sum of squares error function. This approach improves on the linear ranking approach by extending the dependency of state x_n to the outbound and not just the inbound references of the node n. It also describes the output as a function so that the result is a computation that can be applied to a range of inputs rather than a single numerical outcome and allows us to specialize the procedure.

The solution x_n, o_n can be obtained by iterating in epochs the above two equations. Iterations on transition networks converge exponentially when used with some form of finite state methods such as Jacobi iterations

Knowledge graphs

Unlike tightly coupled complex software, deployments manifest more number and variety even as they are isolated from one another. For the most part, deployment stamps follow a template in their construction and deployment but retain dependencies between their instances in their usage. Such a sparse and scattered manifestation of deployed cloud resources finds parallels in other domains where a graph is used to illustrate the "natural relationships". Resource graphs, by that virtue, are popular in the public cloud with cloud providers offering their own flavors and querying language. As with all graphs, two concepts serve our analysis for understanding infrastructure as a graph for an overview. First, is the partition of deployments in terms of the resources in the resource graph either with the help of hierarchy such as resource groups and naming conventions or with the help of tags and attributes of the resources. Second is the abstraction in terms of higher-level graphs that hide the complexity of listings to include only those with higher centrality among all. This applies to both complex pseudo-entities formed from the aggregation of resources as well as simple individual resources of a homogeneous type. In fact, one usage of this graph overview of resources could be detect over-use and high-dependency resources via a complexity metric and builds redundancy into the deployments which also attempts to alleviate load, increase availability and provide business continuity and disaster recovery implementations.

Graphs, although useful to organize and understand the realm of resources across deployments, often demand their own querying language when saved in proprietary databases. Fortunately, the public cloud such as Azure offers a way to include the standard query operators so familiar to the folks from the traditional SQL world along with a way to capture and share these frequently used queries. In addition, the public cloud empowers organizations to connect heterogeneous data stores both in the cloud and on-premises to participate in resource graph queries.

Partitioning and abstractions are left to the organizations to implement and the primary way to do this has been tagging. Some of the tags can even be progressive and representatives of states such as actively used, in-migration, and deprecated to indicate to their resource group, its age and progression towards eventual completion of lifetime scoped by the resource group.

While organization in terms of a resource graph, has been the emphasis for an overview of complex deployments by an entire company, queries have often been analytical in nature and for the benefit of management. Connections in terms of people, places and events elude these queries and management is often required to delegate the creation of new queries, dashboards and reports based on new and evolving questions from encountered incidents. However, a better approach is found to be an investment in customized and meaningful queries from monitoring activities that can be used in conjunction with sophisticated analysis via data mining algorithms and neural network models to continuously provide a backdrop of the health of deployments so that management is no longer reactive from expensive incidents and post-mortem triages but use helpful recommendations and pattern discovery techniques from these queries on a continual basis.

The Hidden Factor

Software CI/CD pipelines authors often miss out on a critical component when it comes to automating IaC deployments simply because they don't have anything equivalent when building applications and databases. IaC providers like Terraform use a concept called state to manage and track the resources they create and manage. The state is a snapshot of the current configuration and status of the infrastructure. The status is usually a json file named say terraform.tfstate that is easy to locate and even documented well but its role escapes attention to those coming from traditional CI/CD pipelines. Modern software engineering including infrastructure deployment is difficult to imagine without Continuous Integration/Continuous Delivery pipelines that are designed to automate and streamline the process of integrating code changes, testing, and deploying code whether the code is for an application or the associated infrastructure. It consists of two practices. The first, Continuous Integration, is one where code changes are merged into a common repository usually multiple times a day and verified by an automated set of build and tests to detect integration issues and improve software quality. The second, Continuous Deployment, is another which builds on the first to automate the release process. It ensures that the software can be released to production at any time with minimal intervention. Together these two practices ensure that capabilities can be added incrementally in small changes that can be independently tested, merged and deployed allowing the end-users to have a snapshot of the code that is always ready to use. The state file goes a step further by allowing the infrastructure provider aka compiler to come up with an execution plan that articulates the necessary incremental changes to the infrastructure. Since the state has persisted as a file, the new configuration and status of the resources can participate in a version control system to track changes and collaborate with others. Another advantage is that the state allows for state locking which prevents concurrent operations from modifying the state simultaneously and this avoids conflicts and inconsistencies.

Finally, regular backups of the state file can prevent data loss and are immensely valuable for recovery and restoration.

The trio of portal, state and IaC must be kept in sync otherwise one of the most perplexing errors that appear is that the changes pushed through the pipeline break unrelated resources.

This section suggests how these three components must be maintained.

Priority:

1. Keep the IaC and state in sync with portal without touching resources.
2. The pipeline must not show conflicts for unrelated changes, edit state.
3. Follow up on any state edits with changes to IaC for resources impacted.

Severity:

1. Maintain associations when adding subnets or virtual networks, allow access to related resources.
2. When versions increase occur, please include them in the portal, state, and code.

Best Practice:

1. Add optional attributes to IaC
2. Prevent unrelated changes to not see conflict.
3. Follow up on any state edits such as version bump or increase count with IaC
4. Keep the planning and apply stages to show similar or no conflicts.

Process:

1. Forward write-through –
   a. Create new resources – complete all associations.
   b. Introduce the state of the new resources.
   c. Create the resources in the portal.
   d. Indicate blockers or announce your changes, when important.
2. Backward propagate changes from Portal
   a. Capture the changes in state
   b. Capture the changes in IaC
   c. Go through step 1 to check that it is no-op
3. Establish baseline and make incremental updates where after each update all three are in sync
4. Add enforcements, detect changes, and send notifications when things change

Finally, the changes being made to keep all three in sync were often spread out over time and distributed among authors, leading to sources of errors or discrepancies. Establishing a baseline combination of state, IaC and corresponding resources is necessary to make changes incremental that is also transparent and quality driven. Keeping them in sync and establishing a timeline with the help of a state file, allows for easier troubleshooting and recovery. The best way to use state files would be to close the gaps by enumerating all discrepancies which also establishes a baseline and then have the process and the practice to ensure that they do not get out of sync.

IaC caveats

These are common errors encountered during the authoring and deployment of Infrastructure-as-Code aka IaC artifacts and these are called out along with their resolutions.

First, resources might pass the identifier of one to another by virtue of one being created before the other and in some cases, these identifiers might not exist during compile time. For example, the code that requires to assign a RBAC based on the managed identity of another resource might not have it during compile time and only find it when it is created during execution time. The RBAC IaC will require a principal _id for which the managed identity of the resource created is required. This might require two passes of the execution – one to generate the RBAC principal id and another to generate the role assignment with that principal id.

The above works for newly created resources with two passes but it is still broken for existing resources that might not have an associated managed identity and the RBAC IaC tries to apply a principal id when it is empty. In such cases, no matter how many times the role-assignment is applied, it will fail due to the incorrect principal id. In this case, the workaround is to check for the existence of the principal id before it is applied.

A second type of case occurs when the application requires an IP address to be assigned for explaining the elaborate firewall rules required based on IP address value rather than references and the IP address is provisioned in the portal before the IaC is applied. This IaC then requires importing the existing pre-created IP address into the state so that the IaC and the state match.

Third, there may be objects in the Key Vault that were created as part of the prerequisites for the IaC deployment and now their ids need to be reconciled with the IaC. Again, the import of that resource into the state would help the IaC provider to reconcile the actual with the expected resource.

Fourth, the friendly names are often references to actual resources that may have long been dereferenced, orphaned, changed, expired, or even deleted. The friendly names, also called keys, are just references and hold value to the author in a particular context but the same author might not guarantee that the moniker is in fact consistently used unless there are some validations and review involved.

Fifth, there are always three stages between design and deploy of Infrastructure-as-code which are "init", "plan" and "apply" and they are distinct. Success in one stage does not guarantee success in the other stage, especially holding true between plan and apply stages. Another limitation is that the plan can be easily validated on the development machine but the apply stage can be performed only as part of pipeline jobs in commercial deployments. The workaround is to scope it down or target a different environment for applying.

Sixth, the ordering and sequence can only be partially manifested with corresponding attributes to explain dependencies between resources. Even if resources are self-descriptive, combination of resources must be carefully put together by the system for a deterministic outcome.

These are only some of the articulations for the carefulness required for developing and deploying IaC.

IaC Challenges

One of the challenges of working with IaC that is somewhat unique to IaC is that authors frequently encounter errors in the 'apply' stage of the IaC and do not detect any errors in the 'plan' stage of the IaC. This leads to write-once-and-fix-many-times and appears to be unavoidable. The compiler only catches a limited set of errors such as when a key is specified instead of an id but whether it is only at runtime can an id be tried and found to be correct or not. A GUID for a principal id is common for role assignments but whether the GUID is appropriate for a particular role assignment depends on the principal to which the GUID belongs as well as the intended target. One way to overcome this limitation is to have a pre-production environment where the code can be applied in a similar way. By the nature of the non-production environment maintaining a separate set of resources than the production environment, sometimes, even this is difficult to do. In such cases, some experimentation might be involved where the IaC is applied once to add and again to remove leaving behind a clean slate. Both non-production and production environments are secured with DevOps pipelines so that IaC is pushed to these environments which results in raising a request and following through each time. Fortunately, there is a better way to scope down problematic or suspicious IaC code snippets and try it out in a personal azure subscription. This approach strongly eliminates all doubts and works without the touch points required for pipelines. And since the sandbox is of no concern to the business, it is even facilitated by organizations to work for all employees and by public cloud as free accounts.

Another challenge that routinely requires more experimentation is for applying permissions to managed identities. Every resource can have its own system managed identity but deployments comprising of resources and their dependencies can have a common user managed identity to govern them. In this case, the identity must be granted permission on all those resources. Several built-in roles varying per resource are applicable to the environment, but the principle of least privileges can only be honored by increasing privileges step-by-step. This calls for a gradation in built-in roles to be tried out for successful application deployments.

Similarly, access is also about connectivity, and it might be surprising that 404 https status code can also imply network failure when the error is being translated from an upstream resource. Some resources have mutual exclusivity between public access and private access. Granting public access with restrictions to some IP addresses might be a hybrid approach that works sufficiently enough to secure resources. It is also important to note that Azure services can bypass general deny rules.

IaC Resolutions

As a recap, almost all IaC providers try to keep pace with the depth and breadth of resource types in the public clouds as well as new features being added to a resource type and while the format of the template can vary between say Azure Resource Manager and Terraform, the IaC provider is usually the resource provider as well or the cloud resource provider also works directly with the IaC provider on building and maintaining related definitions and declarations. This partnership to describe and code cloud resources has gained widespread acceptance and usage, so much so that it can be safely assumed that any capability added to the public cloud at any time can also be captured in IaC code.

Some features including the sought-after preview features are delayed from inclusion in the templates until General Acceptance, but they should still be available from the resource templates from the cloud providers. In such cases, leveraging mixed templates in the IaC source helps to bridge the gap between what can be defined in the IaC and what can be made available from the public cloud. The drift between the public cloud and the IaC is eventually closed.

The folder structure will separate the Terraform templates into a folder called 'module' and the ARM Templates will be located in another folder at the same level and named something like 'subscription-deployments' and include native blueprints and templates. The GitHub workflow definitions will leverage proper handling of either location or trigger the workflow on any changes to either of these locations.

Finally, PowerShell scripts can help with both the deployment as well as the pipeline automations. There are a few caveats with scripts because the general preference is for declarative and idempotent IaC rather than script whether those attributes are harder to enforce, and the logic quickly expands to cover a lot more than originally anticipated. All scripts can be stored in folders with names ending with 'scripts.

It is preferable not to save state in the IaC source code repository and if necessary, it can be stored in the public cloud itself.

With Azure templates, one or more Azure resources can be described with a document, but it does not exist natively in Azure and must be stored locally or in source control. Once those resources are deployed, there is no active connection or relationship to the template.

Terraform and Azure Blueprints are IaC providers that track real-world resources, making operations easier and more powerful. Azure Blueprints preserve the relationship between what should be deployed and what was deployed, allowing improved tracking and auditing. The choice between Azure Blueprint and Terraform depends on their organization methodology, with Azure Blueprints being ideal for compliance and regulations, and Terraform being preferred for flexibility. However, challenges must be addressed, such as centralizing the account with which IaC is deployed and the secrets required for correct deployments. Packaging and distributing solutions for end-users are easier when these can be read from a single source of truth in the cloud, ensuring the location of the solution is known beforehand.

The organization can make use of the best of both worlds with a folder structure that separates the Terraform templates into a folder called 'module' and the ARM Templates in another folder at the same level and named something like 'subscription-deployment' and includes native blueprints and templates. The GitHub workflow definitions will leverage proper handling of either location or trigger the workflow on any changes to either of these locations.

Finally, PowerShell scripts can help with both the deployment as well as the pipeline automations. There are a few caveats with scripts because the general preference is for declarative and idempotent IaC rather than script whether those attributes are harder to enforce, and the logic quickly expands to cover a lot more than originally anticipated. All scripts can be stored in folders with names ending with 'scripts.

IaC is an agreement between the IaC provider and the resource provider. An attribute of a resource can only be applied when the IaC-provider applies it in the way the resource provider expects and the resource-provider provisions in the way that the IaC provider expects. In many cases, this is honored but some attributes can get out of sync resulting in unsuccessful deployments of what might seem to be correct declarations.

For instance, some attributes of a resource can be specified via the IaC provider but go completely ignored by the resource provider. If there are two attributes that can be specified, the resource-provider reserves the right to prioritize one over the other. Even when a resource attribute is correctly specified, the resource provider could mandate the destruction of existing resource and the creation of a new resource. A more common case is one where the IaC wants to add a new property for all resources of a specific resource type but there are already existing resources that do not have that property initialized. In such a case, the applying of the IaC change to add a new property will fail for existing instances but succeed for the new instances. Only by running the IaC twice, once to detect the missing property for the existing resources and initialize and second to correctly report the new property, will the IaC start succeeding in subsequent deployments.

The destruction of an existing resource and the creation of a new resource is also required to keep the state in sync with the IaC. If the resource is missing from the state, it might be interpreted as a resource that was not there in the IaC to begin with and require the destroy before the IaC recognized creation occurs.

It is possible to make use of the best of both worlds with a folder structure that separates the Terraform templates into a folder called 'module' and the resource provider templates in another folder at the same level and named something like 'subscription-deployments' which includes native blueprints and templates. The GitHub workflow definitions will leverage proper handling of either location or trigger the workflow on any changes to either of these locations.

The native support for extensibility depends on naming and logic.

Naming is facilitated with canned prefixes/suffixes and dynamic random strings to make each rollout independent of the previous. Some examples include:

```
resource "random_string" "unique" {
    count = var.enable_static_website && var.enable_cdn_profile ? 1 : 0
    length = 8
    special = false
    upper = false
}
```

Logic can be written out with PowerShell for Azure public cloud which is the de facto standard for automation language. Then a pseudo resource can be added using this logic as follows:

```
resource "null_resource" "add_custom_domain" {
    count = var.custom_domain_name != null ? 1 : 0
    triggers = {always_run = timestamp()}
    depends_on = [
    azurerm_app_service.web-app
    ]

    provisioner "local-exec" {
        command = "pwsh ${path.module}/Setup-AzCdnCustomDomain.ps1"
        environment = {
            CUSTOM_DOMAIN   = var.custom_domain_name
            RG_NAME         = var.resource_group_name
```

```
    FRIENDLY_NAME      = var.friendly_name
    STATIC_CDN_PROFILE = var.cdn_profile_name
    }
  }
}
```

IaC customizations

One of the greatest advantages of using IaC is its elaborate description of the customizations to a resource type for widespread use within an organization. Among the reasons for customization, company policy enforcement, security and consistency can be called out as the main ones. For example, an Azure ML workspace might require some features to be allowed and others to be disallowed before other members of the organization can start making use of it.

There are several ways to do this. Both ARM Templates and azure cli commands come with directives to turn off features. In fact, the 'az feature' command line options are available on a provider-by-provider basis to register or unregister specific features. This ability is helpful to separate out the experimental from the production feature set and allows them to be made available independently. Plenty of documentation around the commands and the listing of all such features makes it easy to work on one or more of them directly.

Another option is to display all the configuration corresponding to a resource once it has been selectively turned on from the portal. Since each resource type comes with its own ARM templates as well as command set, it is easy to leverage the built-in 'show' command to list all properties of the resource type and then edit some of those properties for a different deployment by virtue of the 'update' command. Even if all the properties are not listed for a resource, it is possible to find them in the documentation or by querying many instances of the same resource type.

A third option is to list the operations available on the resource and then set up role-based access control limiting some of those. This approach is favored because users can be switched between roles without affecting the resource or its deployment. It also works for groups and users can be added or removed from both groups and roles. Listing the operations enumerates the associated permissions all of which begin with the provider as the prefix. This list is thorough and covers all aspects of working with the resources. The custom-role is described in terms of permitted 'actions', 'data-actions' and 'not-actions' where the first two correspond to control and data plane associated actions and the last one corresponds to deny set that takes precedence over control and data plane actions. By appropriately selecting the necessary action privileges and listing them under a specific category without the categories overlapping, we create the custom role with just the minimum number of privileges needed to complete a set of selected tasks.

Last but not the least approach, is to supply an initializing script with the associated resource, so that as other users start using it, the initializing script will set the pre-decided configuration with which they must work. This allows for some degree of control on sub resources and associated containers necessary for an action to complete so that by virtue of removing those resources, an action even if permitted by a role on a resource type, may not be permitted on a specific resource.

IaC is not an immutable asset once it is properly authored. It must be maintained just as any other source code asset. Part of the improvements come from fixes to defects, design changes but in the case of IaC specifically, there are other changes coming from drift detection and cloud security posture management aka CSPM.

The separation of workflows from resources and built-to-scale design is a pattern that makes both workflows and resources equally affordable to customers.

IaC for ML Pipelines

Machine Learning experiments are emerging as some of the most popular deployments on the cloud infrastructure where large language models can be easily run on data that is also stored in the cloud. Most machine learning deployment patterns comprise of two types – online inference and batch inference. Both demonstrate MLOps principles and best practices when developing, deploying, and monitoring machine learning models at scale. Development and deployment are distinct from one another and although the model may be containerized and retrieved for execution during deployment, it can be developed independent of how it is deployed. This separates the concerns for the development of the model from the requirements to address the online and batch workloads. Regardless of the technology stack and the underlying resources used during these two phases; typically, they are created in the public cloud; this distinction serves the needs of the model as well.

For example, developing and training a model might require significant computing but not so much as when executing it for predictions and outlier detections, activities that are characteristics of production environments. Even the workloads that make use of the model might vary even from one batch processing stack to another and not just between batch and online processing but the common operations of collecting MELT data, named after metrics, events, logs and traces telemetry and associated resources will stay the same. These include GitHub repository, Azure Active Directory, cost management dashboards, Key Vaults, and in this case, Azure Monitor. Resources and the practice associated with them for the purposes of security and performance are being left out of this discussion, and the standard DevOps guides from the public cloud providers call them out.

Online workloads targeting the model via API calls will usually require the model to be hosted in a container and exposed via API management services. Batch workloads, on the other hand, require an orchestration tool to co-ordinate the jobs consuming the model. Within the deployment phase, it is a usual practice to host more than one environment such as stage and production – both of which are served by CI/CD pipelines that flows the model from development to its usage. A manual approval is required to advance the model from the stage to the production environment. A well-developed model is usually a composite handling three distinct model activities – handling the prediction, determining the data drift in features, and determining outliers in the features. Mature MLOps also includes processes for explainability, performance profiling, versioning and pipeline automations and such others. Depending on the resources used for DevOps and the environment, typical artifacts would include dockerfiles, templates and manifests.

While parts of the solution for this MLOps can be internalized by studios and launch platforms, organizations like to invest in specific compute, storage, and networking for their needs. Databricks/Kubernetes, Azure ML workspaces and such are used for compute, storage accounts and datastores are used for storage, and diversified subnets are used for networking. Outbound internet connectivity from the code hosted and executed in MLOps is usually not required but it can be provisioned with the addition of a NAT gateway within the subnet where it is hosted or with the help of routing tables.

The case for an ML Pipeline

Public clouds lead the way in standardizing deployment, monitoring, and operations for machine learning deployments. Not all development teams are empowered to transition to public cloud because the costs of usage are difficult to articulate upfront of the management and the billing is based on pay-as-you-go model. A Continuous Integration / Continuous Deployment (CI/CD) pipeline, ML tests and model tuning become a responsibility for the development team even though they are folded into the business service team for faster turn-around time to deploy artificial intelligence models in production. In-house automation and development of Machine Learning pipelines and monitoring systems does not compare to those from the public clouds which make it easier for automation and programmability. Yet, transition to public cloud ML pipeline from in-house solution lags due to the following reasons:

First, ML pipeline is a newer technology as compared to traditional software development stacks and management advise that developers have more freedom to explore options on-premises with less cost. Even high-technology large companies with significant investments in hybrid cloud and their own datacenters argue against the use of public cloud technologies. This is not merely from a business point of view; it is also founded with the technical reason that in-house solutions will be better customized to the ML model developments those companies are looking for. Also, experimentation can get out of control from the limits allowed for free-tier. The cost is not always clear, and it always comes down to an argument about the justification of numbers for both options, but the cost is considered lower in favor of the hybrid cloud.

Second, event processing systems such as Apache Spark and Kafka find it easier to replace Extract-Transform-Load solutions that proliferate with data warehouse. It is true that much of the training data for ML pipelines comes from a data warehouse and ETL worsened data duplication and drift making it necessary to add workarounds in business logic. With a cleaner event driven system, it becomes easier to migrate to immutable data, write-once business logic and real-time data processing systems. Event processing systems is easier to develop on-premises even as staging before it is attempted to be deployed to cloud.

Third, Machine learning models are end-products. They can be hosted in a variety of environments, not just the cloud. Some ML users would like to load the model into client applications including those on mobile devices. The model as a service option is rather narrow and does not have to be made available over the internet in all cases especially when the network hop is going to be costly to real-time processing systems. Many IoT traffic and experts agree that the streaming data from edge devices can be quite heavy in traffic where an online on-premises system will out-perform any public-cloud option. Internet tcp relays are of the order of 250-300 milliseconds whereas the ingestion rate for real-time analysis can be upwards of thousands of events per second.

When it comes to picking a cloud resource for developing Machine Learning Pipelines, there are certain prerequisites that stand out such as a comprehensive toolset, collaboration and version control, automations, integrations with other services, security and compliance. Azure Machine Learning is one such resource which provides an environment to create and manage the end-to-end life cycle of Machine Learning models. Machine Learning's compatibility with open-source frameworks and platforms like PyTorch and TensorFlow makes it an effective all-in-one platform for integrating and handling data and models which tremendously relieves the onus on the business to develop new capabilities. Azure Machine Learning is designed for all skill levels, with advanced MLOps features and simple no-code model creation and deployment.

Azure Machine Learning has a drag and drop interface that can be used to train and deploy models. It uses a machine learning workspace to organize shared resources such as pipelines, datasets, compute resources, registered models, published pipelines, and real-time endpoints. A visual canvas helps build end to end machine learning workflow. It trains, tests, and deploys models all in the designer. The datasets and components can be dragged and dropped onto the canvas. A pipeline draft connects the components. A pipeline run can be submitted using the resources in the workspace. The training pipelines can be converted to inference pipelines and the pipelines can be published to submit a new pipeline that can be run with different parameters and datasets. A training pipeline can be reused for different models and a batch inference pipeline can be used to make predictions on new data.

The steps to create a machine learning pipeline in Azure Machine Learning begins with the creation of a workspace. It serves as the central hub for managing all machine learning dependencies. The next step requires setting up data stores which allow access to data needed for pipelines. By default, each workspace has a default data store connected to a storage account for saving snapshots of work done but additional datastores that encapsulate connection parameters and credentials to external data stores can also be registered. Each pipeline manifests an intention, so it's helpful to break down the task into manageable stages such as data preparation, model training and evaluation. With the rich programmability options in terms of language-based Software Development Kits, these can be articulated and decorated with labels to indicate that they are automations to be run. Each such script will require a compute target where it can be run. Whether it is an individual virtual machine or a cluster of nodes to run on depends on whether the automation is zero-touch or not. There must also be orchestration for the pipeline to manage dependencies between the steps or for preamble and epilogue and the order in which the steps must be run along with error handling and retry behaviors. When the pipeline is ready, it must be publish-able and shared for use. Finally, there must be monitoring and tracking for the health of the components involved, to detect data drift and for remedial actions.

We will compare this environment with some open-source options such as TensorFlow but for those unfamiliar with the latter, here is a use case. A JavaScript application performs image processing with a machine learning algorithm. When enough training data images have been processed, the model learns the characteristics of the drawings which results in their labels. Then as it runs through the test data set, it can predict the label of the drawing using the model. TensorFlow has a library called Keras which can help author the model and deploy it to an environment such as Colab where the model can be trained on a GPU. Once the training is done, the model can be loaded and run anywhere else including a browser. The power of TensorFlow is in its ability to load the model and make predictions in the browser itself.

The labeling of drawings starts with a sample of say a hundred classes. The data for each class is available on Google Cloud as numpy arrays with several images numbering say N, for that class. The dataset is pre-processed for training where it is converted to batches and outputs the probabilities.

As with any ML learning example, the data is split into 70% training set and 30% test set. There is no order for the data and the split is taken over a random set.

TensorFlow makes it easy to construct this model using the TensorFlow Lite ModelMaker. It can only present the output after the model is trained. In this case, the model must be run after the training data has labels assigned. This might be done by hand. The model works better with fewer parameters. It might contain 3 convolutional layers and 2 dense layers. The pooling size is specified for each of the convolutional layers, and they are stacked up on the model. The model is trained using the tf.train.AdamOptimizer() and compiled with a loss function, optimizer just created, and a metric such as top k in terms of categorical accuracy. The summary of the model can be printed for viewing the model. With a set of epochs and batches, the model can be trained. Annotations help TensorFlow Lite converter to fuse TF.Text API. This fusion leads to a significant speedup than conventional models. The architecture for the model is also tweaked to include projection layer along with the usual convolutional layer and attention encoder mechanism which achieves similar accuracy but with much smaller model size. There is native support for HashTables for NLP models.

With the model and training/test sets defined, it is now easy to evaluate the model and run the inference. The model can also be saved and restored. It is executed faster when there is GPU added to the computing.

When the model is trained, it can be done in batches of predefined size. The number of passes of the entire training dataset called epochs can also be set up front. A batch size of 256 and the number of steps as 5 could be used. These are called model tuning parameters. Every model has a speed, Mean Average Precision, and output. The higher the precision, the lower the speed. It is helpful to visualize the training with the help of a high chart that updates the chart with the loss after each epoch. Usually there will be a downward trend in the loss which is referred to as the model is converging.

When the model is trained, it might take a lot of time, say about 4 hours. When the test data has been evaluated, the model's efficiency can be predicted using precision and recall, terms that are used to refer to positive inferences by the model and those that were indeed positive within those inferences.

The pipeline can also be used to evaluate models, like for example, with Machine Learning Studio. Foundation models can be evaluated with proprietary test data. Microsoft developed foundation models are a great way to get started with data analysis on your data. The Model Catalog is the hub for both foundation models as well as OpenAI models. It can be used to discover, evaluate, fine tune, deploy and import models. Your own test data can be used to evaluate these models. The model card on any of the foundational models can be used to pass in the test data, map the columns for the input data, based on the schema needed for the task, provide a compute-instance to run the evaluation on, and submit the job. The results include evaluation metrics, and these can help decide if you would like to fine tune the model using your own training data. Every pre-trained model from the model catalog can be fine-tuned for a specific set of tasks such as text classification, token classification, and question answering. The data can be in JSONL, CSV, or TSV format and the steps are just like evaluation except that you will pass in validation data for validation and test data to evaluate the fine-tuned model. Once the models are evaluated and fine-tuned, they can be deployed to endpoints for inferencing. There must be enough quota available for deployment.

OpenAI models differ from the foundation models in that they require a connection with Azure OpenAI. The process of evaluating, fine-tuning, and deploying remains the same. An Azure Machine Learning Pipeline can be used to complete a machine learning task which usually consists of three steps: prepare data, train a model, and score the model. The pipeline optimizes the workflow with speed, portability, and reuse so you can focus on machine-learning instead of infrastructure and automation. A pipeline comprises of components for each of the three tasks and is built using the Python SDK v2, CLI or UI. All the necessary libraries such as azure.identity, azure.ai.ml, and azure.ai.ml.dsl can be imported. A component is a self-contained piece of code that does one step in a machine learning pipeline. For each component, we need to prepare the following: prepare the python script containing the execution logic, define the interface of the component, and add other metadata of the component. The interface is defined with the "@command_component" decorative to Python functions. The studio UI displays the pipeline as a graph and the components as blocks. The input_data, training_data, and test_data are the ports of the component which connect to other components for data streaming. Training and scoring are defined with their respective Python functions. The components can also be imported into the code. Once all the components and input data are loaded, they can be composed into a pipeline.

The Azure Machine Learning Studio allows us to view the pipeline graph, check its output and debug it. The logs and outputs of each component are available to study them. Optionally components can be registered to the workspace so they can be shared and reused.

A pipeline component can also be deployed as a batch endpoint. This is helpful to run machine learning pipelines from other platforms such as custom Java code, Azure DevOps, GitHub Actions, and Azure Data Factory. A Batch endpoint serves REST API so it can be invoked from other platforms. By isolating the pipeline component as a batch endpoint, we can change the logic of the pipeline without affecting downstream consumers. A pipeline must first be converted to a pipeline component before being deployed as a batch endpoint. Time-based schedules can be used to take care of routine jobs. A schedule associates a job with a trigger which can be cron.

Workflows

Workflows are built upon the artifacts in a tiered architecture where different workflows can make use of the same artifacts differently. There are several workflow management software that can be used for infrastructure deployment purposes.

For example, Apache AirFlow is a platform used to build and run workflows. A workflow is represented as a Directed Acyclic Graph where the nodes are the tasks, and the edges are the dependencies. This helps to determine the order in which to run them and with retries. The tasks are self-described. An Airflow deployment consists of a scheduler to trigger scheduled workflows and to submit tasks to the executor to run, an executor to run the tasks, a web server for a management interface, a folder for the DAG artifacts, and a metadata database to store state. The workflows do not restrict what can be specified as a task which can be an Operator or a predefined task using say Python, a Sensor which is entirely about waiting for an external event to happen, and a Custom task that can be specified via a Python function decorated with an @ task label.

Runs of the tasks in a workflow can occur repeatedly by processing the DAG and can occur in parallel. Edges can be modified by setting the upstream and downstream for a task and its dependency. Data can be passed between tasks using an XCom, a cross-communications system for exchanging states, uploading, and downloading from an external storage, or via implicit exchanges.

Airflows send out tasks to run on workers as space become available so they can fail but they will eventually complete. Notions for sub-DAGs and TaskGroups are introduced for better manageability.

One of the characteristics of AirFlow is that it prioritizes flow, so that there is no need to describe data input or output, and all aspects of the flow can be visualized whether they include pipeline dependencies, progress, logs, code, tasks, and success status.

AirFlow is in use by over ten thousand organizations with popular use cases involving orchestrating batch ETL jobs, organizing, executing, and monitoring data flow, building ETL pipelines for extracting batch data from hybrid data sources and running Spark jobs, training machine learning models, generating automated reports, and performing backups and other DevOps tasks. It might not be ideal for streaming events because the scheduling required is different between batch and stream. Also, AirFlow does not offer versioning of pipelines, so source control might become necessary for such cases. AirFlow epitomizes pipelines as a code with artifacts described in Python for creating jobs, stitching jobs, programming other necessary data pipelines and debugging and troubleshooting.

Workflows tend to lower the emphasis on an actor and her actions and instead provide a trackable mechanism where multiple actors can pitch in. It is a different way of thinking and execution, so the tenets and best practices are also different from the organization world of teams and charters. These are:

- Reusability – many of the activity from the library of activities for one workflow can and will be reused for another. Very few workflows might have differences in doing tasks that were not covered by the global collection of activities. There should not be any difference between an activity that appears in bootstrapping and its invocation during redeployment/ rehosting in the new environment. Only the parameter values will change for this.

- Dependencies – many of the dependencies will be implicit as they originate from system components and services information. A workflow might additionally specify dependencies via the standard way in which workflows indicate dependencies. These will be on a case-by-case basis for tenants since it adds overhead to other services, many of whom are standalone. Implicit dependencies can be articulated in the format specified by the involved components.

- Splitting – Workflows are written for on-demand invocation from the web interface or by the system, so there might be more than one for a specific deployment scenario. It is best to include both the bootstrapping and the redeploy in the main workflow for the specific scenario, but they will be mutually exclusive during their respective phases and remain idempotent.

- Idempotency – All workflow steps and activities should be idempotent. If there are conditionals involved, they must be part of activities. The signaling and receiving notifications of dependent workflows if any must be specifically called out.

- Bootstrapping – This phase is common to many services and usually requires at least a cluster/set of servers to be made ready but there might be activities that require the service stamp to be deployed even if it is not configured along with necessary activities to do one time preparation such as getting secrets. Until the VIPs are ready, the redeployment cannot be kicked off. Bootstrapping might involve preparations for both primary and secondary where applicable.

- Redeployment or rehosting – This phase involves configuration since the bootstrapping is usually for a stamp and this stage converts it into a deployment for a service. Since it involves reconfiguration, it can be for both primary and secondary and typically done inside the new cloud. It is best to parameterize as much as possible.

- Naming convention – Though workflows can have any names inside the package that the owning teams upload, it is best to follow a convention for the specific scenario of one workflow calling another. Standalone single workflows do not have this problem. Even in the case when there are many workflows, a prefix/suffix might be helpful. This applies to both work workflows and activities.

- System workflow – Requiring separate workflows for bootstrap and redeployment via a system defined workflow to allow system to inject system defined activities between bootstrap and redeploy is a nice-to-have but the less intrusion into service deployment the better. This calls on the service to do their own tracking via passing parameter values between workflows and activities. A standard need not be specified for this, and it can be left to the discretion of the services.

The above list is not intended to be complete but focuses on the strengths of those that have worked well.

Comparisons to Deis Workflow

While workflows can be imagined for any level and scope of impact, they can also be articulated on different systems. While public cloud based systems are inherently described by IaC, on-premises systems are often based on state reconciliation systems like Kubernetes. Deis workflow is a platform-as-a-service that adds a developer friendly layer to any Kubernetes cluster so that applications can be deployed and managed easily. Kubernetes evolved as an industry effort from the native Linux container in support of the operating system. It can be considered as a step towards a truly container centric development environment. Containers are lightweight resources with just enough memory and compute partitioning to host code. Containers decouple applications from infrastructure as a layer in between which separates development (Dev) from operations (Ops).

Containers made Platform-as-a-Service aka PaaS possible. Containers help compile the code for isolation. PaaS enables applications and containers to run independently. PaaS containers were not open source. They were just proprietary to PaaS. This changed the model towards the development of container centric frameworks where applications could now be written with their own.

Let us look at the components of the Deis workflow:

The workflow manager – checks your cluster for the latest stable components. If the components are missing. It is essentially a Workflow Doctor providing first aid to your Kubernetes cluster that requires servicing.

The monitoring subsystem consists of three components – the Telegraf, InfluxDB, and Grafana. The first is a metrics collection agent that runs using the daemon set API. The second is a database that stores the metrics collected by the first. The third is a graphing application, which natively stores the second as a data source and provides a robust engine for creating dashboards on top of the time-series data.

The logging subsystem which consists of two components – first that handles log shipping and second that maintains a ring buffer of application logs

The router component which is based on Nginx and routes inbound https traffic to applications. This includes a cloud-based load balancer automatically.

The registry component which holds the application images generated from the builder component.

The object storage component where the data that needs to be stored is persisted. This is generally an off-cluster object storage.

Slugrunner is the component responsible for executing build-pack based applications. Slug is sent from the controller which helps the Slugrunner download the application slug and launch the application

The builder component is the workhorse that builds your code after it is pushed from source control.

The database component which holds most of the platform state. It is typically a relational database. The backup files are pushed to object storage. Data is not lost between backup and database restarts.

The controller serves as the http endpoint for the overall services so that CLI and SDK plugins can be utilized.

Deis Workflow is more than just an application deployment workflow examples of which include entities like CloudFoundry that provide a multi-cloud multi-language abstraction layer for infrastructure to support scalability and DevOps integration and especially when they want convenience over Kubernetes. Deis Workflow performs application rollbacks, supports zero-time app migrations at the router level and provides scheduler tag support that determines which nodes the workloads are scheduled on. Moreover, it runs on Kubernetes so other workloads can be run on Kubernetes along with these workflows. Workflow components have a "deis-" namespace that tells them apart from other Kubernetes workloads and provide building, logging, release and rollback, authentication and routing functionalities all exposed via a REST API. This is a layer distinct from the Kubernetes as well as CloudFoundry. While Deis provides workflows, Kubernetes provides orchestration and scheduling.

Workflows can be authored in the public cloud such as Azure using resources for DevOps. The usage patterns remain the same although the syntax and semantics might vary. For example, projects can be added to support different business units. Within a project, teams can be added. Repositories and branches can be added for a team to empower them with the necessary tools. Agents, agent pools, and deployment pools can be added to support continuous integration and deployment for the teams contributions. Azure Active Directory becomes the central registry to manage users and groups for membership to teams.

Solution Accelerator

One of the recent infrastructure trends has been the use of solution accelerator. This is a collection of resources with a configuration that is best suited to a workload. These pre-configured, canned, and ready-to-go deployment templates are the equivalent of prefabricated items in any industry. They could range in size from small storage to large, range in purpose from supporting a dedicated to workload to all kinds of applications, and range in cost from least expensive to one with all bells and whistles. These solution accelerators can be published by home-grown teams as well as imported from other organizations and their repositories. In this section, we will cover only the case where a solution accelerator must be developed in-house.

One of the main characteristics of a solution accelerator is that it is intended to be paired with a workload. Understanding the workload and implementing the best practices into a deployment stamp so that its blueprint can be used many times, is a prerequisite. It is this reusability of the solution accelerators that drives it to take the form of T-Shirt sizes so that workloads of the same type both big and small can leverage the same solution accelerator.

Another characteristic of the solution accelerator is that it must be built like a product and the fewer and more suited to the demand the better. When we treat it like a product rather than a solution where one is more generic and the other more specific, we become more focused in our efforts. This does not compromise the business purpose for which the solution accelerator is designed. It merely reduces waste in terms of unnecessary distractions, prototypes, and their iterations.

As such the following principles apply to a solution accelerator: 1. Know Your Customers: Understanding your customers' needs, behaviors, and pain points is crucial. This involves continuous user research and feedback collection. 2. Outcome Over Output: Focus on the outcomes and impact of the whole proposal rather than just the features. This means prioritizing features that deliver the most value to users. 3. Prioritization: Effective prioritization is essential. Solution Accelerators developers must balance various factors such as customer needs, business goals, and technical feasibility. 4. Data-Driven Decision Making: Leverage data and analytics to inform decisions. This helps in understanding user behavior, measuring success, and making informed adjustments. 5. Cross-Functional Collaboration: Work closely with engineering, design, marketing, and other teams. Effective communication and collaboration are key to aligning everyone towards common goals. 6. Agility and Adaptability: Be prepared to pivot and adapt based on new information, market changes, or feedback. Flexibility is crucial in the fast-paced tech industry. 7. Clear Vision and Roadmap: Develop a clear solution accelerator vision and roadmap. This helps in setting expectations, aligning the team, and guiding the development process. 8. Experimentation and Innovation: Encourage a culture of experimentation. Testing new ideas and iterating based on results can lead to innovative solutions. 9. Customer-Centric Mindset: Always keep the customer at the center of decision-making. This ensures that the solution accelerator remains relevant and valuable to its users. 10. Effective Communication: Ensure transparent and effective communication within the team and with stakeholders. This helps in managing expectations and keeping everyone informed.

These principles help solution accelerator developers navigate the complexities of the infrastructure and deliver a blueprint that meets the user needs and drives business success.

IaC Modernization

Software Applications evolve as business requirements change and to pay off technical debt. While Application Modernization is eased by a cloud recognized adoption framework, strategy, business, and even dedicated cloud services, IaC transformations are not talked about in the same way. Hosts for a software application and their IaC are also veritable assets that need to keep up with the modernization of the digital assets of an organization. The intention in this section is to draw some parallels and propose some automations for the 'modernization' of IaC.

Figure depicts the modernization across both applications (above) and hosting infrastructure (below).

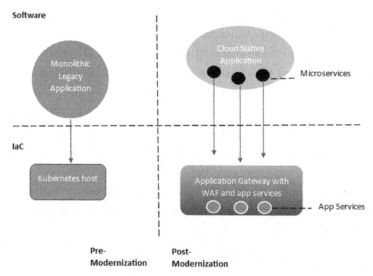

Figure also illustrates that the lifetime of an infrastructure is scoped to the code and data it hosts.

Software used by companies is critical to their business and will continue to provide a return on investment. Companies will try to maximize this for as long as possible. Some maintenance is required of these software systems which satisfy business and customer needs and address technical debt that accrues over time. Maintenance works well for short-term needs but as time progresses, the systems become increasingly complex and out of date. Eventually maintenance will no longer be efficient or cost-effective. At this point, modernization is required to improve the system's maintainability, performance, and business value. It takes much more effort to accomplish compared to maintenance. If software can no longer be maintained or modernized, it will need to be replaced.

The risks of modernizing legacy systems primarily come from missing documentation. Legacy systems seldom have complete documentation specifying the whole system with all its functions and use cases. In most cases, the documentation is missing badly, which makes it hard to rewrite a system that would function identically to the previous one. Companies usually couple their legacy software with their business processes. Changing legacy software can cause unpredictable consequences to the business processes that rely on it. The nature of replacing legacy systems with new ones is risky since the new system can be more expensive on a total cost of ownership basis and there can be problems with its schedule of delivery.

There are at least three strategies for dealing with legacy systems: scrap the legacy system, keep maintaining the system or replace the whole system. Companies generally have limited budgets on the legacy systems, so they want to get the best return on the investment. Scrapping the system can be an option if the value has diminished sufficiently. Maintenance can be opted into when it is cost-effective. Some improvement is possible by adding new interfaces to make the system easier to maintain. Replacement can be attempted when the support has gone, the maintenance is too expensive, and the cost of the new system is not too high.

Both technical and business perspectives are involved. If a legacy system has low quality and low business value, the system should be removed. Those with low quality but high business value must be maintained or modernized depending on the expense. Systems with high quality can be left running.

Modernization is a more extensive process than maintenance because modernization often incorporates restructuring, functional changes, and new software attributes. Modernization can be either white-box or black-box depending on the level of abstraction. White box modernization requires a lot of information about the internals of the legacy system. Contrary to that, the black box modernization only requires external interfaces and compatibility. Replacement is an option when neither approach works.

Software modernization is also an evolution of systems. White box systems are more popular than black box systems which might be counter-intuitive to the notion that black-box modernization is easier than white-box modernization. The tool for whitebox methods could have become better to help with the shift. Legacy systems are harder to integrate. Software integration allows companies to better control their resources, remove duplicate business rules, re-use existing software, and reduce the cost of development. The effort needed to keep legacy systems running often takes resources away from other projects. Legacy systems also suffer from diminishing ownership and knowledge base which makes changes difficult to make.

Infrastructure Modernization is not always a one-to-one mapping between software and their hosts. They undergo considerations that are specific to the infrastructure plane and include factors such as grouping, visibility, choice of hosting resources and deployment templates, sizing and scaling, scope, and lifetimes. Every new host for a component of an erstwhile application comes with additional features and settings that must be specified requiring more IaC than before and often more complex than before as the hosting becomes more cloud native. In fact, pre-modernization IaC is similar to on-premises application in terms of technical debt.

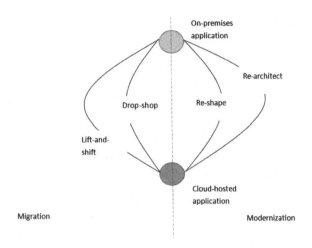

Fig 2. Illustrates the migration versus modernization differences.

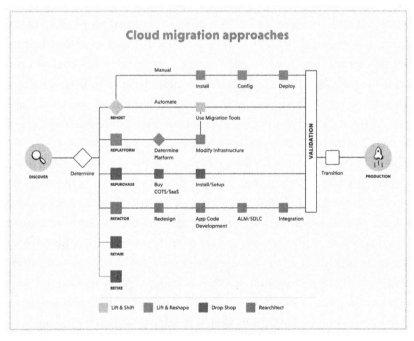

Fig 3. Shows all the cloud migration approaches.

The refactoring of old code to new microservices

Among the strategies called out for cloud migration approaches, refactoring holds the most promise for flexibility and relevancy. With the introduction of microservices, it became easy to breakout monolithic code into independent technology stacks for dedicated business purposes and separate the concerns for each functionality. Not only were these independently testable but the user interface could mash up one or more backend microservices to render the pages. There were no restrictions to the hosts for the microservices. One could even use Mesos-based clusters and shared volumes instead of virtual machines or containers which was a significant win for high availability and failover. This is possibly great for small and segregated data, but larger companies often require massive investments in their data, often standardizing tools, processes, and workflows to better manage their data independent of the business verticals they served. In such cases, consumers of the data do not talk to the database directly but via a service that sits behind even say a message bus that could fulfill requests with data from different data sources and provide asynchronous I/O pattern. If the consumers proliferate, they end up creating and sharing many different instances of services for the same data, each with its own view rather than the actual table. APIs for these services are more domain-based than implementing a query-friendly interface that lets you directly work with the data. As services are organized into a flat ring of independent services or one behind the other for data access, data may get translated or massaged as it makes its way from one to another. Data may even be at most one or two fields of an entity along with its identifier for such services. This works very well to alleviate the onus and rigidity that comes with organization, the interactions between the components, and the various chores that need to be taken to keep it flexible to suit changing business needs. The microservices are independent so they stand by themselves as if spreading out from data for their respective functionalities. This is already business-friendly because each service can now be modified and tested independently of others.

The transition to microservices from legacy monolithic code is not straightforward. The functionalities must be separated beyond components. And in the process of doing so, we cannot risk regression. Tests become a way to scope out behavior at boundaries such as interface and class interactions. Adequate coverage of tests will guarantee backward compatibility for the system as it is refactored. The microservices are independently testable both in terms of unit tests as well as end-to-end tests. Services usually have a REST interface which makes it easy to invoke them from clients and comes with the benefits of using browser-based developer tools. The data store does not need to be divided between services. In some cases, only a data access service is required which other microservices can call. The choice and design of microservices stem from the minimal functionalities that need to be separated and articulated. If the services do not need to be refactored at a finer level, they can remain encapsulated in a singleton.

The rule of thumb for the refactoring of the code is the follow-up of the Do not Repeat Yourself or (DRY) principle which is defined as "Every piece of knowledge must have a single, unambiguous, authoritative representation within a system." This calls for every algorithm or logic that is cut and pasted for different usages to be consolidated at a single point of maintenance. This improves flexibility because enhancements such as the use of a new data structure can be replaced in one place, and it also reduces the bugs that come by when similar changes must be made in several places. This principle also reduces the code when it is refactored, especially if the old code had several duplications. It provides a way to view the minimal skeleton of the microservices when aimed at the appropriate scope and breadth. Even inter-service calls can be reduced with this principle.

Good microservices are not only easy to discover from their APIs but also easy to read from their documentation which can be autogenerated from the code with markdowns. Different tools are available for this purpose and both the approach of using microservices as well as the enhanced comments describing the APIs provide sufficient information for the documentation.

Promoting cloud services to cloud-native infrastructure gives more leeway for those app services to evolve in the future with little or no infrastructure changes.

Maintainability, performance, and security

The maintainability of microservices is somewhat different from conventional software. When the software is finished, it is handed over to the maintenance team. This model is not favored for microservices. Instead, a common practice for microservices development is for the owning team to continue owning it for its lifecycle. Amazon's "you build it, you run it" philosophy inspires this idea. Developers working daily with their software and communicating with their customers create a feedback loop for the improvement of the microservice.

Microservices suffer a weakness in their performance in that communication happens over a network. Microservices often send requests to one another. The performance is dependent on these external request-responses. If a microservice has well-defined bounded contexts, it will experience less performance hits. The issues related to microservice connectivity can be mitigated in two ways – making less frequent and more batched calls as well as converting the calls to be asynchronous. Parallel requests can be issued for asynchronous calls and the performance hit is that of the slowest call.

Microservices have the same security vulnerabilities as any other distributed software. Microservices can always be targeted for denial-of-service attacks. Some endpoint protection, rate limits and retries can be included with the microservices. Requests and responses can be encrypted so that the data is never clear. If the "east-west" security cannot be guaranteed, at least the edge facing microservices must be protected with a firewall or a proxy or a load balancer or some such combination. East-West security refers to the notion that the land connects the east and the west whereas the oceans are external. Another significant security concern is that monolithic software can be broken down into many microservices which can increase the surface area significantly. It is best to perform threat modeling of each microservice independently. Threat modeling can be done with STRIDE as an example. It is an acronym for the following: Spoofing Identity – is the threat when a user can impersonate another user. Tampering with data- is the threat when a user can access resources or modify the contents of security artifacts. Repudiation – is the threat when a user can perform an illegal action that the microservice cannot deter. Information Disclosure – is the threat when, say, a guest user can access resources as if the guest was the owner. Denial of service – is the threat when say a crucial component in the operations of the microservice is overwhelmed by requests so that others experience outage. Elevation of privilege – is the threat when the user has gained access to the components within the trust boundary and the system is therefore compromised.

Migration of microservices comes with three challenges: multitenancy, statefulness and data consistency. The best way to address these challenges involves removing statefulness from migrated legacy code, implementing multitenancy, and paying increased attention to data consistency.

Infrastructure used post modernization of applications also has similar challenges and must use proper choices and settings for the cloud resources.

Path towards Microservices architecture

The path towards a microservice-based architecture is anything but straightforward in many companies. There are plenty of challenges to address from both technical and organizational perspectives. The performed activities and the challenges faced during the migration process are both included in this section.

The migration to microservices is sometimes referred to as the "horseshoe model" comprising three steps: reverse engineering, architectural transformations, and forward engineering. The system before the migration is the pre-existing system. The system after the migration is the new system. The transitions between the pre-existing system and the new system can be described via pre-existing architecture and microservices architecture.

The reverse engineering step comprises the analysis by means of code analysis tools or some existing documentation and identifies the legacy elements which are candidates for transformation to services. The transformation step involves the restructuring of the pre-existing architecture into a microservice based one by reshaping the design elements, restructuring the architecture, and altering business models and business strategies. Finally, in the forward engineering step, the design of the new system is finalized.

Many companies will say that they are in the early stages of the migration process because the number and size of legacy elements in their software portfolio continues to be a challenge to get through. That said, these companies also deploy anywhere from a handful to hundreds of microservices while still going through the deployment. Some migrations require several months and even a couple of years. The management is usually supportive of migrations. The business-IT alignment comprising technical solutions and business strategies are more overwhelmingly supportive of migrations.

Microservices are implemented as small services by small teams that suits Amazon's definition of the Two-Pizza Team. The migration activities begin with an understanding of both the low-level and the high-level sources of information. The source code and test suites comprise the low-level. The higher-level comprises of textual documents, architectural documents, data models or schema and box and lines diagrams. The relevant knowledge about the system also resides with people and in some extreme cases as tribal knowledge. Less common but useful sources of information include UML diagrams, contracts with customers, architecture recovery tools for information extraction and performance data. Very rarely but also found are cases where the pre-existing system is considered so bad that its owners do not look at the source code.

Such an understanding can also be used towards determining whether it is better to implement new functionalities in the pre-existing system or in the new system. This could also help with improving documentation, or for understanding what to keep or what to discard in the new system. Change in the application architecture post-modernization warrants a revamping of the hosting infrastructure and not just a decomposition of the existing hosts. This calls for opportunities to consolidate storage, computing, and networking in favor of potential growth of the new architecture components.

Application Migration Process

There are at least two ways that are frequently encountered for the migration process itself. In some cases, the migration towards microservices architecture is organized in small increments, rather than a big overall migration project. In those cases, migration is implemented as an iterative and incremental process. They might also be referred to as phased adoption. This has been the practice even for the migration towards Service Oriented Architecture. There are times when the migration has a predefined starting point but not necessarily a defined upfront endpoint.

Agility is a very relevant aspect when moving towards a microservices architecture. New functionalities are often added during the migration. This clearly shows that the preexisting system was hindering development and improvements. New functionalities are added as microservices, and existing functionalities are reimplemented also as microservices. The difficulty is only in getting the infrastructure ready for adding microservices. Domain-driven design practices can certainly help here.

Not all the existing functionality is migrated. It does not align with the "hide the internal implementation detail" principle of microservices nor does it align with the typical MSA characteristic of decentralized data management. If the data is not migrated, it may hinder the evolution of independent services. Both the service and the data scalability are also hindered. If scalability is not a concern, then the data migration can be avoided altogether.

The main challenges in architecture transformation are represented by (i) the high level of coupling, (ii) the difficulties in identifying the boundaries of services (iii) and system decomposition. There could be some more improvement and visibility in this area with the use of architecture recovery tools so that the services are well-defined at the architectural level.

Some good examples of microservices have consistently shown a pattern of following the "model around business concepts."

The general rule of thumb inferred from various microservices continues to be 1) First, to build and share reusable technical competence/knowledge which includes (i.) kickstarting a MSA and (ii.) reusing solutions, 2) Second, to check business-IT alignment which is a key concern during the migration and 3) Third, to monitor the development effort and migrate when it grows too much which would show a high correlation between migration to microservices and increasingly prohibitive effort in implementing new functionalities in the monolith.

IaC transformations for purely rehosting changes are much more straightforward than modernization. The new hosts will likely preserve the control and data planes.

Incremental code migration strategy

Let us review a case study on the incremental code-migration strategy of large monolithic base used in supply system. The code migration strategy considers a set of factors that includes scaffolding code, balancing iterations, and grouping related functionality.

Incremental migration only works when it is progressive. Care must be taken to ensure that progress is measured by means of some key indicators. These include tests, percentage of code migration, signoffs, and such other indicators. Correspondingly the backlog of the code in the legacy system that must be migrated must also decrease.

Since modernized components are being deployed prior to the completion of the entire system, it is necessary to combine elements from the legacy system with the modernized components to maintain the existing functionality during the development period. Adapters and other wrapping techniques may be needed to provide a communication mechanism between the legacy system and the modernized systems when dependencies exist.

There is no downtime during the incremental modernization approach and this kind of modernization effort tries to always keep the system fully operational while reducing the amount of rework and technical risk during the modernization. One way to overcome challenges in this regard is to plan the efforts involved in modernization. A modernization plan must also include the order in which the functionality is going to be modernized.

Another way to overcome challenges is to build and use adapters, bridges and other scaffolding code which represents an added expense, as this code must be designed, developed, tested, and maintained during the development period but it eventually reduces the overall development and deployment costs.

Supporting an aggressive and yet predictable schedule also helps in this regard. The componentization strategy should seek to minimize the time required to develop and deploy the modernized system.

This does not necessarily have a tradeoff with quality as both the interim and final stages must be tested and the gates for release and progression towards revisions only help with the overall predictability and timeline of the new system.

Risk occurs in different forms and some risk is acceptable if it is managed and mitigated properly. Due to the overall size and investment required to complete a system migration, it is important that the overall risk be kept low. The expectations around the system, including its performance, helps to mitigate these risks.

IaC development can also keep up with the phases of deployments needed to mirror the code migration strategy. This provides opportunities to automate repeatable deployments.

Towards architecture driven modernization

The paths to modernization aren't always a matter of business priority and available tools. The science behind modernization involves meta modeling and analyzing transformations of models that can help to optimize costs. This is done in three phases: reverse engineering, restructuring and forward engineering. Reverse engineering technologies can analyze legacy software systems, identify their widgets and their interconnection, reproduce it based on the extracted information, and create a representation at a higher level of abstraction. Some requirements for modernization tools can be called out here. It must allow extracting domain classes according to Concrete Syntax Tree meta-model and semantic graphical information, then analyze extracted information to change them into a higher level of abstraction as a knowledge-discovery model.

Software modernization approach becomes a necessity for creating new business value from legacy applications. Modernization tools are required to extract a model from text in source code that conforms to a grammar by manipulating the concrete syntax-tree of the source code. For example, there is a tool that can convert Java swing applications to Android platform which uses two Object Management Group standards: Abstract Syntax Tree for representing data extracted from java swing code in reverse engineering phase and Knowledge Discovery Platform-independent model. Some tools can go further to propose a Rich Internet Application Graphical User Interface. The three phases articulated by this tool can be separated into stages as: the reverse engineering phase which uses the JDTAPI for parsing the Java swing code to fill in an AST and Graphical User Interface model and the restructuring phase that represents a model transformation for generating an abstract KDM model and the forward phase which includes the elaboration of the target model and a Graphical User Interface.

The overall process can be described as the following transitions:

Legacy system
–parsing–>
AST Meta model
–restructuring algorithm–>
Abstract Knowledge Model
--forward engineering–>

GUI Metamodel.

The reverse engineering phase is dedicated to the extraction and representation of information. It defines the first phase of reengineering following the Architecture Driven Modernization process. It is about the parsing technique and representing information in the form of a model. Parsing can focus on the structural aspect of header and source files and then there is the presentation layer that determines the layout of the functionalities such as widgets.

The restructuring phase aims at deriving an enriched conceptual technology independent specification of the legacy system in a knowledge model KDM from the information stored inside the models generated on the previous phase. KDM is an OMG standard and can involve up to four layers: Infrastructure layer, Program Elements layer, resource layer and abstractions layer. Each layer is dedicated to a particular application viewpoint.

The forward engineering is a process of moving from high-level abstractions by means of transformational techniques to automatically obtain representation on a new platform such as microservices or as constructs in a programming language such as interfaces and classes. Even the user interface can go through forward engineering into a Rich Internet application model with a new representation describing the organization and positioning of widgets.

Automation is key to developing a tool that enables these transitions via reverse engineering, restructuring and forward engineering. IaC transformations are also subject to automation. A knowledge model of the existing IaC is just as useful as the knowledge for the existing software prior to modernization. Reverse and forward engineering are so rampant for infrastructure development in different industries that they deserve elaboration.

Reverse engineering using models

Software models are not just great for learning about software, but they are essential to software modernization, which usually consists of three steps: reverse engineering, restructuring and forward engineering. When extracted from legacy software, models are easier to transform into a different architecture such as microservices. This activity could be performed by tools to bridge grammarware and model-driven technical spaces. Such dedicated parsers could also come with a query language that eases the retrieval of scattered information in syntax trees. It could also incorporate extensibility and grammar reuse mechanisms.

Model driven software development raises the level of abstraction and automation in the construction of software. Although commonly used for building new software systems, models also have the potential to evolve existing systems. They help to reduce the software evolution costs and improve the quality of the artifacts. Migration and modernization can both benefit from model driven approaches. A set of standard metamodels have been popularized by solutions and tools which represent the information normally managed in modernization tasks. This is true for language-to-language migrations as well where the first step is to extract models from the application code written in source language. Similarly, in a modernization effort, models could be extracted from schemas to improve the data quality. When these initial models are obtained, transformations can then be applied to generate higher level abstraction models followed by generation of artifacts such as code in another language or an improved data schema. Other operations on models such as model comparisons and synchronizations can also be applied.

The relationship between pairs of concepts grammar or program and metamodel or model is an example of a bridge between two different technical spaces specifically grammarware and modelware. Dedicated parsers are implemented to obtain models from code conforming to a grammar. These parsers perform model generation tasks in addition to code parsing. First, a syntax tree is created from the source code by static analysis and then this syntax tree is traversed to obtain the information needed to create the model elements. This is a complex task which requires both collecting scattered information and resolving references in the syntax tree.

Model transformations are classified into three categories which include text to model transformations which obtain models from existing source code; model-to-model transformations whose input and output are models; and models-to-text transformations, which generate software artifacts from a source model. When a model transformation language is designed, two key design choices are how to express the mappings between source and target elements and how to navigate through the source artifact. The former involves binding, and the latter involves querying.

Certain improvements can be imagined with the use of 1) reuse mechanisms at rule-level 2) a new kind of rule for dealing with expressions efficiently, and 3) an extensibility mechanism to add new operators. New functionalities could also target expressiveness, usability, and performance. Reverse-engineering of the IaC can proceed in parallel with that for the software application.

Model-driven Software development

Model driven Software development evolves existing systems and facilitates the creation of new software systems.

The salient features of model-driven software development include:

1. Domain-specific languages (DSLs) that express models at different abstraction levels.
2. DSL notation syntaxes that are collected separately
3. Model transformations for generating code from models either directly by model-to-text transformations or indirectly by intermediate model-to-model transformations.

An abstract syntax is defined by a metamodel that uses a metamodeling language to describe a set of concepts and their relationships. These languages use object-oriented constructs to build metamodels. The relationship between a model and a metamodel can be described by a "conforms-to" relationship.

There are seven metamodels including Knowledge Discovery Metamodel, Abstract Syntax Tree Metamodel, the Software Measurement Metamodel, analysis program, visualization, refactoring, and transformation

ASTM and KDM are complimentary in modeling software systems' syntax and semantics. ASTMs use Abstract Syntax Trees to mainly represent the source code's syntax, KDM helps to represent semantic information about a software system, ranging from source code to higher levels of abstractions. KDM is the language of architecture and provides a common interchange format intended for representing software assets and tools interoperability. Platform, user interface or data can each have its own KDM and are organized as packages. These packages are grouped into four abstract layers to improve modularity and separation of concerns: infrastructure, program elements, runtime resource and abstractions.

SMM is a metamodel that can represent both metrics and measurements. It includes a set of elements to describe the metrics in KDM models and their measurements.

Taking the example of the modernization of a database forms application and migrating it to a Java platform, an important part of the migration could involve PL/SQL triggers in legacy Forms code. In a Forms application, the sets of SQL statements corresponding to triggers are tightly coupled to the User Interface. The cost of the migration project is proportional to the number and complexity of these couplings. The reverse engineering process involves extracting KDM models from the SQL code.

An extractor that generates the KDM model from SQL code can be automated. A framework that provides domain specific languages for extraction of model is available and this can be used to create a model that conforms to a target KDM from program that conforms to grammar. Dedicated parsers can help with this code-to-model transformation.

With the popularity of machine learning techniques and SoftMax classification, extracting domain classes according to syntax tree metamodel and semantic graphical information has become more meaningful. The two-step process of parsing to yield Abstract Syntax Tree Metamodel and restructure to express Abstract Knowledge Discovery Model becomes enhanced with collocation and dependency information. This results in classifications at code organization units that were previously omitted. For example, code organization and call graphs can be used for such learning as shown in reference 1. The discovery of KDM and SMM can also be broken down into independent learning mechanisms with the Dependency Complexity being one of them.

KDM helps to represent semantic information about a software system, ranging from source code to higher level of abstractions. KDM is the language of architecture and provides a common interchange format intended for representing software assets and tools interoperability. Platform, user interface or data can each have its own KDM and are organized as packages. These packages are grouped into four abstract layers to improve modularity and separation of concerns: infrastructure, program elements, runtime resource and abstractions.

SMM is a metamodel that can represent both metrics and measurements. It includes a set of elements to describe the metrics in KDM models and their measurements.

The Architecture Driven Modernization process comprises of two main steps: the Knowledge Discovery Metamodel (KDM) model extraction and metric report generation. The process can be walked through in this manner. The source code is converted into an Abstract Syntax Tree model using a Code-to-Model transformation. The Abstract Syntax Tree model is converted into a KDM model using a model-to-model transformation. The KDM model converts to a metrics model using another model-to-model transformation. Finally, a metrics report is generated from the metrics model using a model-to-text transformation.

If we take the example of a set of SQL statements converted to Pl/SQL Abstract Syntax Tree Metamodel, it will consist of definitions like RDBTableDefinition, RDBColumnDefinition and such others, primitive types which consist of RDBTableType, RDBColumnType, RDBDatabaseType and such others, statements that comprise RDBSelectStatement, RDBModifyStatement, RDBInsertStatement and such others, BinaryExpressions such as RDBSelectExpression, RDBHostVariableExpression and such others.

Extracting KDM models from source code in a general-purpose programming language (GPL) is a key activity in the application of Abstract Data Model performed during the reverse engineering phase of the application modernization. The overall process consists of reverse engineering, restructuring and forward engineering.

When the models are extracted from the GPL code, the main task is collecting scattered information for creating the model elements from source code. The scattering occurs due to the references between the elements. When such references are explicit in the models, they are implicitly established in the source code with the use of identifiers such as the reference between a variable and its declaration. Transforming an identifier-based reference into an explicit reference involves looking for the identified element in the source code. Dedicated parsers result from this challenge. This scattering problem requires complex processing to locate the correspondence between source code and the model elements. A powerful XPath-like language specially built for resolving references can help here.

KDM models are generated ASTM models by applying a chain of M2M transformations. First, an initial transformation generates an L0 KDM model from the ASTM model. Then we can convert the generated model from one or more L1 KDM models, depending on which architectural views of the legacy system are needed to achieve the desired results.

Once the KDM models are generated, there is an appropriate representation of the existing system to generate artifacts related to the modernization process. With the specific example of modernizing database forms that involve SQL triggers, several metrics can be defined to measure the coupling that influences the efforts of migrating triggers. For example, these metrics are based on the UI statements' count, location, and type such as whether for reading or writing. The couplings can be classified as reflective, declarative, and imperative. The extracted KDM models can then be transformed into Software Measurement Metamodels.

With the popularity of machine learning techniques and SoftMax classification, extracting domain classes according to syntax tree meta-model and semantic graphical information has become more meaningful. The two-step process of parsing to yield Abstract Syntax Tree Meta-model and restructure to express Abstract Knowledge Discovery Model becomes enhanced with collocation and dependency information. This results in classifications at code organization units that were previously omitted. For example, code organization and call graphs can be used for such learning as shown in reference 1. The discovery of KDM and SMM can also be broken down into independent learning mechanisms with the Dependency Complexity being one of them.

Dependencies are not limited to the application plane and have similar manifestations in the infrastructure plane. Corresponding efforts taken independently on the infrastructure plane can yield significant and similar benefits.

Rehost, Refactor, Rearchitect, Rebuild, Retire

One of the Gartner Reports for Information Technology called out the paths to adopting the public cloud as one of Five R's – Rehost, Refactor, Rearchitect, Rebuild and Retire. Many Solution Architects would tend to agree with such separate workstreams for the progression from assessment through deployment and finally release of the cloud hosted applications. But while the analysts and even customers tend to focus on one approach, the Solution architects often see many paths and even involving intermediary steps between the start and the finish. While they fine tune the path by optimizing several parameters and often on a case-by-case basis, they generally agree with the break-out of these definitions.

Rehosting is a recommendation to change the infrastructure configuration of the application to "Lift and Shift" it to the cloud using Infrastructure-as-a-service.

Refactor is a recommendation to perform modest modifications of the application code without changing the architecture or functionality so that it can be migrated to the cloud in a container using Container-as-a-service or Platform-as-a-service.

Rearchitect is a recommendation to dramatically modify the application code thereby altering the architecture to improve the health of the application and enable it to be migrated to the cloud using Platform-as-a-service or deployed serverless using Function-as-a-service.

Rebuild is a recommendation to discard the code of the application and develop it again in the cloud using Platform-as-a-service or serverless Function-as-a-service.

Retire is a recommendation to discard the application altogether or potentially replace it with a commercial software-as-a-service alternative.

All of the recommendations above can be made on an application-by-application basis but must be elaborated with additional information such as the Lines of code analyzed, the number of files read, a score or ranking, the total number of full-time employees required, a quantitative assessment for cloud readiness, the number of roadblocks encountered, the estimated effort, Operations and Business Support system dedication, the SR, SA, and SE assessments, and tentative dates for the release.

As you can see, this process can be quite intensive and repeating it for dozens of legacy applications can be quite tedious. Producing a table for the results of the rigor across all these applications and for a dashboard to represent the insights via canned queries can easily become error prone. Fortunately, some trail and documentation are maintained as this analysis is conducted and there are quite a few tools that can help with the visualization of the data even as the data is incrementally added to the dataset. An end-to-end automation that can help with detailed analysis, drill downs and articulated charts and graphs is possible if those applications were merely scanned by their source. Indeed, a rapid scan of a portfolio of applications can be a great start in the roadmap that requires us to assess, mobilize, migrate, and modernize but iterations will remain inevitable and even desirable as they help towards optimizing continuously. Visualization of infrastructure transformations are veritable assets after the migration as they provide insights that were never explored earlier.

A recommendation to businesses

Small and medium businesses often prioritize capabilities and assets that add value for the customer by counting applications and unfortunately sideline infrastructure into costs. Even though the cloud onboards these businesses with a clear roadmap and self-serve migration paths, they are forced to develop the applications and infrastructure themselves. Often this means that they minimize infrastructure footprint and maximize application development that is well-within their domain. Even though infrastructure is a commodity, the businesses are best served by clubbing application development and infrastructure operations together and ensuring that the infrastructure suits the applications like a glove in the hand. Many vendors and solution integrators step in to fill the gap between the business and technical expectations. These businesses and vendors are best served by a pattern for software development and testing in the cloud that lets them develop their applications rapidly and at scale with minimal overhead. Public clouds even offer lightweight infrastructure in the form of serverless computing for these purposes, but discipline and pattern are required by the implementors. With the elimination of a one-time deployment of migrated code, the path to incremental delivery and enhancement of logic that is hosted in the cloud becomes more gradual, deliberate, and even allows outreach to be expanded independent of development. Versioning of the logic helps keep existing customers while onboarding new ones. The practice of "DevOps" is a testament to this agile development required by these businesses to shorten the time to build and deliver capabilities from what used to take weeks and months with systems to hours and days with cloud services.

Let us take a specific example of an open-source Linux-based web application developed with NodeJS, saved on GitHub and built and deployed with a pipeline such as AWS CodeBuild. The steps described here help to set up a continuous integration and continuous delivery workflow that runs unit-tests from a GitHub repository. Unit-tests reduce refactoring time while helping engineers to get up to speed on their code base more quickly and provide confidence in the expected behavior. It involves testing individual functions including Lambda functions. Use of an AWS Cloud9 instance which is an Integrated Development Environment is suggested but not mandatory for these implementors, even though it can be accessed even through a web browser. The idea here is to emphasize that the experience is well thought-through and facilitated with tools to follow a groove.

Setting up this pipeline involves a few milestones which developers call epics. The first epic involves running unit-tests on a personal GitHub repository with CodeBuild. The tasks involved are:

Sign into the AWS Management Console and open the CodeBuild console.

Create a build project and in the project configuration, type the name of the project

In the source section, specify the provider as GitHub and point to the existing personal Repository in the GitHub account by specifying its URL.

In the primary source webhook events section, specify rebuilding every time a code change is pushed to this repository.

In the environment section, choose managed image and the latest image for an Amazon Linux instance.

Leave the default settings and complete the project creation.

Start the build.

The CodeBuild console will display the tests run and the unit-test results can be reviewed. These results validate the repository integration with the project that has been created with the steps above. When the webhook is applied, code changes automatically start a build.

Unit-tests often involve assertion, spies, stubs and mocks.

An assertion is used to verify an expected result. For example, the following code validates that the results are in a given range:

```
Describe('Target Function Group', () => {
It('Check that the result is between 0 and 1000', function () {
        const target = new Target();
        expect(target.id).is.above(0).but.below(1000)
    });
});
```

A spy is used to observe what is happening when a function is running. The following example shows whether a set of methods were invoked.

```
Describe('Target Function Group', () => {
It ('should verify that the proper methods were called', () => {
        const spyStart = spy(Target.prototype, "start");
        const spyStop = spy(Target.prototype, "stop");
        const target = new Target();
        target.start();
        target.stop();
        expect(spyStart.called).to.be.true;
        expect(spyStop.called).to.be.true;
    });
});
```

A stub is used to override a function's default response. For example, a stub can be used to force a return ID from the getId function

```
Describe('Target Function Group', () => {
    It ('Check that the Target Id is between 0 and 1000', function () {
        let generateIdStub = stub(Target.prototype, 'getId').returns(99999);
        const target = new Target();
        expect(target.getId).is.equal(99999);
        generateIdStub.restore();
    });
});
```

A mock is a fake method that has predefined behavior for testing different scenarios. A mock can be considered an extended form of a stub and can carry out multiple tasks simultaneously. For example, a mock can be used to validate three scenarios: 1. A function is called 2. That it is called with arguments. and 3. It returns an integer, say, 9.

```
Describe('Target Function Group', () => {
    It ('Check that the TargetId is between 0 and 1000', function() {
        let mock = mock(Target.prototype).expects('getId').withArgs().returns(9);
        const target = new Target();
        const id = target.getId();
        mock.verify();
        expect(id).is.equal(9);
    });
});
```

Infrastructure unit-testing and integration testing as called out here is just a taste but its significance to the quality of the overall application and infrastructure development cannot be understated. Monitoring and reporting are also required for the overall tracking of health and performance as well as to receive notifications when things go wrong, and these are available via the infrastructure's observability and programmability features.

Field Guide

This section recognizes that application and infrastructure migration is as much art as science. While the theory and practice outlined in earlier sections continue to apply, there is not a single solution that is an exact fit for the problem at hand and it would help to at least hit a home run on the most important asks from a project. Consider this field guide more as a summary of the discussion so far and a caveat for not being sticky about any one aspect.

Application modernization is about reverse engineering, restructuring and forward engineering. It is also an opportunity to optimize the user's experience through modernization. Customers are vital to business. Applications have become the gateway to more impactful and rewarding experiences for internal stakeholders and customers. Modernization is not just one of the ways but sometimes the only way for applications to embrace agility, fuel growth and remain competitive.

The points listed here are all elements of a robust hybrid cloud strategy and are essential for a full modernization experience including applications and infrastructure. They can be used to accelerate digital transformations by building new capabilities and delivering them. Cloud native architectures and containerizations are priorities. Delivery must be accelerated with a culture of automation and transformation and deployments must be friendly to hybrid clouds.

One of the most critical aspects of application modernization is application readiness assessment. A cloud-native microservices approach can bring scalability and flexibility inherent to the cloud but it relies on an evaluation of the existing application. It also brings the opportunity to tailor the application to the business needs.

The build once and deploy on any cloud begins with assessing the applications. Some of them can undergo lift-and-shift while others will require refactoring. Even if the applications are deployed with little changes, preparing them for containerization is essential. Containers bring scalability, openness, and portability. Automating deployments via a CI/CD pipeline is next. DevOps pipelines are very welcome here. Applications must be run and managed with ease for true embrace by customers.

Accelerators and tools can certainly help but recognizing the disciplines in which they will be used are just as important. For example, Innovation helps with refactoring which can help deliver a cloud-native application. Agile delivery can help with "replatform" that when deployed by modern DevOps pipelines and run by newer runtimes can help deliver a cloud ready application. Cost reduction is another distinct area where repackaging can help save costs if a traditional application is delivered. Cloud Migration that requires VMs in the cloud and when used with migration accelerators and common operations can help deliver complex traditional applications in the new world.

There is not a single formula that holds but an approach that is totally unique to the business goals and needs. The modernization goals of agile delivery, transform-and-innovate, reduce costs, replace with SaaS, and cloud migrations can be planned for by analyzing for insights and the utilization of one or more modernization patterns that include pivoting from monolithic strategies, adding new capabilities as microservices, refactoring monolith to microservices, exposing APIs, and migrating monolith to a cloud runtime. With these, applications can be deployed to both a public cloud and a private cloud.

A trusted foundation helps. Infrastructure platforms like Kubernetes and cloud technologies like AWS and Azure developer tools provide a consistent platform to leverage for this purpose. By their availability in the cloud as cloud resources, it is possible to preserve some of the hosting infrastructure, but transformations can better serve the changes in traffic and data access.

Sample Worksheet for application modernization

Application Name	Lines-of-code	Resources	Agility	Flexibilty	TechnicalDebt	#person-months	Recommendations
Application 1	>150K	3	0.2	0.72	0.2	2	Refactor
Application2	>500M	8	0.3	0.13	0.64	8	Rearchitect
Application3	>100K	2	0.4	0.22	0.4	1	Rehost
Application4	<90K	1	0.1	0.45	0.11	0	Lift-and-shift

A similar worksheet for infrastructure modernization will help with similar recommendations in the infrastructure plane except for the lift-and-shift as existing infrastructure is usually locked-in.

Questionnaire

Application Modernization Readiness Assessment
The survey will take approximately 4 minutes to complete.
This checklist evaluates the dependencies of an application to help with its modernization. It strives to collect all the information about the application to give a more detailed and precise picture of its readiness for building and running it in the cloud.

1. The drive for the modernization of this application comes from:
   Multiple choice.
   Changing business requirements
   Technical debt
   Pending deadline
   Budgetary considerations
   None of the above

2. My application is an N-Tier web application and has a customer facing user interface. Single choice.
   Yes
   No

3. My application modernization journey requires all three: planning, executing, and monitoring. Single choice.
Yes
No

4. My application is accessible (Select all that apply.) Multiple choice.
Web Interface
Command line
Scripts
SDK
None of the above

5. I would like my cloud adoption strategy to be Single choice.
Retain/Retire
Lift-and-shift
Lift-and-reshape
Replace, drop and shop
Refactor (Rewriting/De-coupling applications)

6. My application has specific requirements from: (Select all that apply) Multiple choice.
Programming Languages
Operating systems
Databases
Services
Application Frameworks

7. My application must maintain the same programming language Single choice.
Yes
No

8. My application is sensitive to the flavor and/or version of operating system Single choice.
   Yes
   No

9. My application requires the database to be the same as before: Single choice.
   Yes
   No

10. My application is dependent on other services that are not available in the cloud Single choice.
    Yes
    No

11. My application requires profiling to generate: Multiple choice.
    mapping of system components
    topology maps
    coverage of the technology stack
    automations
    creating a baseline
    to view/simulate real-world conditions
    for full stress testing

12. I am fine with phased migration with phases for Multiple choice.
    service-by-service migration
    improving performance and scalability
    Integration and full DevOps support
    meeting SLAs required from the application

13. I can point to SLAs for the application Single choice.
    Yes
    No

14. I need CI/CD enhancements for Multiple choice.
    visibility over migration strategy and roadmap
    investing over quality controls
    showing on dashboards
    using my monitoring solution

15. I need investment in fault detection for Multiple choice.
    maintaining availability and performance
    leveraging my monitoring investment
    improving visibility
    reducing false alerts

16. I have specific queries or demands from my applications behavior over time Single choice.
    Yes
    No

17. The number of web requests to my application are in the range Single choice.
    < 100 per hour
    < 100 per minute
    < 100 per second
    100 - 1000 per second
    > 1000 per second

18. The size of data stored in the database increases by Single choice.
    a few hundred bytes per day
    a few hundred kilobytes per day
    a few hundred megabytes per day
    a few hundred gigabytes per day
    a few terabytes per day
    greater than a few terabytes a day

19. My application data must be encrypted (Select all that apply) Multiple choice.
    at rest
    in transit

20. My data has PHI and is subject to governance and compliance Single choice.
    Yes
    No

21. With the adoption of cloud technologies, the number of users to my application is expected to increase by Single choice.
    < 5%
    5-15%
    15-50%
    > 50%

22. My application depends on an on-premises message broker. The throughput of the queues is in the range Single choice.
    0-1 Kilobytes/second
    1-5 Kilobytes/second
    < 1 MB/second
    None of the above

23. My application has ETL, data pipelining and/or batch automation jobs involved. Single choice.
    Yes
    No

24. My application has vendor-lock ins for OLAP and data warehouses. Single choice.
    Yes
    No

25. My application has disaster recovery considerations and/or involves data ageing and archival. Single choice.
Yes
No

26. Application response times must be in the range of Single choice.
< 100 ms
100 - 250 ms
> 250 ms

27. My data must remain on premises. Single choice.
Yes
No

28. My application has significant security restrictions and makes use of Firewalls, vulnerabilities assessments and periodic threat assessments. Single choice.
Yes
No

29. My application must meet security certifications such as ISO 27001, ISO 27017 (cloud security), ISO 27019 (privacy), ISO 9001, AWS PCI, and SOC 1,2, and 3, HIPAA, FERPA, CJIS, SEC Rule 17a-4(f), IRS 1075, and SRG Impact level 2 and 4 for DoD systems. Single choice.
Yes
No

30. My application must work for the government and must comply with FedRAMP at the Moderate and High Level or the GDPR or PCI-DSS or such others. Single choice.
Yes
No

31. My application must reduce the average time needed to detect an intrusion or security failure (MTTD) or the average time needed to resolve issues such as a security breach or outage (MTTR) Single choice.
Yes
No

32. My application must demonstrate an order of magnitude higher availability after moving to the cloud. Single choice.
Yes
No

33. My application must localize workloads to a specific geographic region. Single choice.
Yes
No

34. I have a specific preference to a specific cloud and I am willing to sacrifice one or more of the following (Select all that apply) Multiple choice.
technological scale and expertise to handle critical and highly complex workloads
anticipated cost savings
internal IT burden
None of the above

The Application Migration scenario for serving static content – a case study

Single Page applications and static websites are quite popular for hosting code because they don't necessarily need a compute or a container to host them and can even be served as files from a variety of distribution points for public access. By their nature, they are extremely portable and can run from filesystems as well as internet open directories. Yet, security and performance are not always properly considered for their deployments. This section delves into an architectural pattern to host static website content in the public cloud.

We chose Amazon AWS as the public cloud for this scenario, but the pattern is universal and applies across most major public clouds.

When static content is hosted on AWS, the recommended approach is to use an S3 bucket as the origin and CloudFront to distribute the content geographically. There are two primary benefits to this solution. The first is the convenience of caching static content at edge locations. The second involves defining web access control lists for the CloudFront distribution. This helps to secure requests to the content with minimal configuration and administrative overhead.

The only limitation to this standard recommended approach is that in some cases, virtual firewall appliances may need to be deployed in a virtual private cloud to inspect all content. The standard approach does not route traffic through the virtual private cloud. Therefore, an alternative is needed that still uses a CloudFront distribution to serve static content in an S3 bucket, but the traffic is routed through the VPC by using an Application Load Balancer. An AWS Lambda function then retrieves and returns the content from the S3 bucket.

The resources in this pattern must be in a single AWS region, but they can be provisioned in different AWS accounts. The limits apply to the maximum request and response size that the Lambda function can receive and send, respectively.

There must be a good balance between performance, scalability, security, and cost-effectiveness when using this approach. While Lambda can scale for high availability, the number of concurrent executions must not exceed the maximum quota otherwise requests will be denied.

The architecture lays out the CloudFront as facing the client and communicating with a firewall and two load balancers – one in each availability zone of the region hosting all these resources. The load balancers are created in the public subnet within a virtual private cloud, and both communicate with a Lambda function that serves content from a private S3 bucket. When the client requests a URL, the CloudFront distribution forwards the request to a firewall which filters the request using the web ACLs applied to the CloudFront distribution. If the request cannot be served from the internal cache within CloudFront, it is forwarded to the load balancer which has a listener associated with the target group based on a Lambda function. When the Lambda function is invoked, it performs a GetObject operation on the S3 bucket and returns the content as a response.

The deployment of the static content can be updated using a Continuous Integration / Continuous Deployment facilitating pipeline.

The Lambda function introduced in this pattern can scale to meet the load from all the load balancers and its security can be tightened by specifying the origin as the S3 bucket and similarly for the distribution to have the origin as the load balancer. Instead of IP, DNS name can be used to refer to the resources.

Well-architected framework

Public clouds provide an adoption framework for businesses that helps to create an overall cloud adoption plan that guides programs and teams in their digital transformation. The plan methodology provides templates to create backlogs and plans to build necessary skills across the teams. It helps rationalize the data estate, prioritize the technical efforts, and identify the data workloads. It is important to adhere to a set of architectural principles which help guide development and optimization of the workloads. A well-architected framework stands on five pillars of architectural excellence which include:

- Reliability (REL)
- Security (SEC)
- Cost Optimization (COST)
- Operational Excellence (OPS)
- Performance efficiency (PERF)

The elements that support these pillars are a review, a cost and optimization advisor, documentation, patterns-support-and-service offers, reference architectures and design principles.

This guidance provides a summary of how these principles apply to the management of the data workloads.

Cost optimization is one of the primary benefits of using the right tool for the right solution. It helps to analyze the spend versus time as well as the effects of scale out and scale up. An advisor can help improve reusability, on-demand scaling, reduced data duplication, among many others.

Performance is usually based on external factors and is very close to customer satisfaction. Continuous telemetry and reactiveness are essential to tuned up performance. The shared environment controls for management and monitoring create alerts, dashboards, and notifications specific to the performance of the workload. Performance considerations include storage and compute abstractions, dynamic scaling, partitioning, storage pruning, enhanced drivers, and multilayer cache.

Operational excellence comes with security and reliability. Security and data management must be built right into the system at layers for every application and workload. The data management and analytics scenario focus on establishing a foundation for security. Although workload specific solutions might be required, the foundation for security is built with the Azure landing zones and managed independently from the workload. Confidentiality and integrity of data including privilege management, data privacy and appropriate controls must be ensured. Network isolation and end-to-end encryption must be implemented. SSO, MFA, conditional access and managed service identities are involved to secure authentication. Separation of concerns between azure control plane and data plane as well as RBAC access control must be used.

The key considerations for reliability are how to detect change and how quickly the operations can be resumed. The existing environment should also include auditing, monitoring, alerting and a notification framework.

In addition to all the above, some consideration may be given to improving individual service level agreements, redundancy of workload specific architecture, and processes for monitoring and notification beyond what is provided by the cloud operations teams.

Each pillar contains questions for which the answers relate to technical and organizational decisions that are not directly related to the features of the software to be deployed. For example, a software that allows people to post comments must honor use cases where some people can write, and others can read. But the system developed must also be safe and sound enough to handle all the traffic and should incur reasonable costs.

Since the most crucial pillars are OPS and SEC, they should never be traded in to get more out of the other pillars.

The security pillar consists of Identity and access management, detective controls, infrastructure protection, data protection and incident response. Three questions are routinely asked for this pillar:

1. How is the access controlled for the serverless api?
2. How are the security boundaries managed for the serverless application?
3. How is the application security implemented for the workload?

The operational excellence pillar is made up of four parts: organization, preparation, operation, and evolution. The questions that drive the decisions for this pillar include:

1. How is the health of the serverless application known?
2. How is the application lifecycle management approached?

The reliability pillar is made of three parts: foundations, change management, and failure management. The questions asked for this pillar include:

1. How are the inbound request rates regulated?
2. How does resiliency build into the serverless application?

The cost optimization pillar consists of five parts: cloud financial management practice, expenditure and usage awareness, cost-effective resources, demand management and resources supply, and optimizations over time. The questions asked about cost optimization include:

1. How are the costs optimized?
2. Are there avenues for consolidation?

The performance efficiency pillar is composed of four parts: selection, review, monitoring and tradeoffs. The questions asked for this pillar include:

1. How is the performance optimized for the serverless application?

In addition to these questions, there is quite a lot of opinionated and even authoritative perspectives on the appropriateness of a framework, and they are often referred to as lenses. With these forms of guidance, a well-architected framework moves closer to reality.

The power of infrastructure consolidation

Infrastructure deployed to the public clouds is often characterized by shared-nothing deployments that are dedicated to different workloads. With the convenience of repeatable and automated deployments via Infrastructure-as-Code aka IaC, it becomes easy to spin up as many resources as necessary to separate the usages. This is advantageous in many ways. For example, deployments can evolve differently and have different lifetimes and growth spurts. Ownership can be handed in to independent teams. Access control can be set up differently. Different workloads can require different tunings which might manifest as resource-level settings and configurations. The deployments can be scaled independently. It is also possible to bill them in different accounts or at least tag them differently. State may be captured in the tags and labels differently between the resources of the same type in those deployments. The explosion in the number of resources is not a concern for automation that repeats the same steps for each. Workloads can be studied more effectively, and they do not share the same fate. Troubleshooting and maintenance become easier and cheaper. It is even possible to establish a control set for baseline.

But consolidation has its own merits. For example, web applications deployed to independent application services can be hosted on the same application service plan when they do not have widely varying requirements. A single app service plan requires only one subnet for virtual network integration to guarantee that all outbound traffic goes via the virtual network. This is especially helpful to ensure a robust private plane connectivity between storage and computing resources that these web applications might need. Cost savings come from avoiding higher SKU app service plans as well as those from associated resources such as subnets and networking appliances. It provides another level of differentiation between infrastructure and workload perspectives albeit at the resource level instead of the resource group level. While simultaneously deployed resources and their dependencies are easy to interpret from the written IaC for deployments, the savings have a ripple effect on other non-functional aspects such as logging and monitoring. With a reduction in maintenance, consolidation also lowers costs for skills, training, and dedication. Consolidation as a strategy is even a prerequisite to creating service tiers so that quality-of-service aka QoS can be better articulated. The resource pooling and workload classification to pools is an established pattern for improving resource utilization in any infrastructure whether it is on-premises and at cloud scale.

Even the savvy cloud account owners are wary of ballooning costs over time for resources and there are many features available natively from the cloud to address some of these concerns, but no cloud can truly understand the resource utilization or refactoring possible without some involvement from these owners. Case in point is the use of specific resource types and monitoring practices to set up feedback-loop cycles. Both AWS and Azure offer resource types such as X-Rays and Application Insights to analyze and debug distributed applications which collects data about the requests that the application serves and provides tools to view, filter, and gain insights into that data. Yet owners seldom take advantage of setting up monitoring and dashboards for all their assets. Infrastructure providers aka IaC deployers are left to fill the gap but they have one advantage that others don't. They can do this for entire resource groups and deployments that are not limited to specific web applications or resource types. Even if the dashboard they create and maintain to study the infrastructure usage patterns become private, they will be empowered to advise the account owners on tactical consolidations. They are not limited to making inferences from cost management dashboards and utilization dashboards and patterns can also help them to validate their choice of SKUs, reservations, and other assertions. They are not strangers to alerts and notifications and resource-level chores, but they could benefit from creative ways to set up deployment scope feedback cycles.

IaC Innovations

The automation for deploying infrastructure in a consistent, repeatable, and error-free manner, involves assets, we call code because it takes similar diligence to be written as any application code that performs create, update, and delete of business entities. Infrastructure-as-code has been devops-oriented, often gaining mutually reinforcing automation between the pipeline and the infrastructure delivered. Technology varies with cloud native forms, providers like Ansible, Terraform, and domain specific language such as Pulumi. IaC can be curated as a set of machine-readable files, descriptive model, and configuration templates. There are two approaches for writing it: an imperative approach and a declarative approach. The imperative approach allows users to specify the exact steps to be taken for a change and the system does not deviate from them while a declarative approach specifies the final form, and the tool or platform involved goes through the motion of provisioning them. As with DevOps based growth, IaC is usually short-sighted often manifesting independent shared-nothing and even redundant deployments for different business scenarios, leaving little room for consolidation, consistency, and organization across deployment hierarchies of management group, subscription, resource groups and resource types.

Yet complex IaC deployment writers invest in many notable routines to bring cloud best practices, and artifact management to IaC. These include uniform, consistent, and global naming conventions, registries that can be published by the system for cross subscription and cross region lookups, parameterizing diligently at every scope including hierarchies, leveraging dependency declarations, and reducing the need for scriptability in favor of system and user defined organizational units of templates. This section introduces an infrastructure knowledge base that can live in the cloud and is continuously updated with each deployment to provide supportability via read-only stores and frequently publishing continuous and up-to-date information on the rollouts and their key performance indicators and help alleviate the operations from the design and development of IaC.

Such a knowledge base would be a hub and spoke model correlating resources with their origin information and existence identifiers and provide a one-stop shop for information regarding the infrastructure that would otherwise require manual cross linking of git repositories, pipeline runs, and issue tracking that are usually disparate systems in themselves. It would make the store cloud native so that the information can be available to query just like the various graph explorers that provide unparalleled querying and automation capabilities for resources and identities in the cloud. This extra mile for a knowledge management system in the cloud as a re-purposable data model across infrastructure maintaining organizations provides more than a single pane of glass for infrastructure drill downs and cross-reference. It brings a platform and enables many virtuous feedback cycles that serve IaC writers and arms them with more information at their fingertips before they make any changes.

Many might point to native supportability from existing tracking systems including issue and code repositories given that files and not databases serve better for difficult to automate clouds such as sovereign clouds, and regions with fewer resource-types availabilities. It is true that the practice of annotating every commit in a repository with rich links to origin, growth and timelines can also provide independent sources of information that can be spanned by custom queries as the need arises, but it remains an extra mile, and most teams are left to fulfil that themselves leading to boutique solutions. On the other hand, incident tracking software alone has demonstrated the effectiveness of a knowledge base that supports ITSM, ITBM, ITOM and CMDB capabilities.

In addition, a realization dawns in, as the size and scale of infrastructure grows that the veritable tenets of IaC such as reproducibility, self-documentation, visibility, error-free, lower TCO, drift prevention, joy of automation, and self-service somewhat diminish when the time and effort increase exponentially to overcome its brittleness. Packages go out of date, features become deprecated and stop working, backward compatibility is hard to maintain, and all existing resource definitions have a shelf-life. Similarly, assumptions are challenged when the cloud provider and the IaC provider describe attributes differently. The information contained in IaC can be hard to summarize in an encompassing review unless we go block by block and without a knowledge base, this costly exercise is often repeated. It is also easy to shoot oneself in the foot by means of a typo or a command during the exercise and especially when the state of the infrastructure disagrees with that of the portal.

The data model would articulate Infrastructure-as-a-code and blueprints, resources, policies, and accesses as an entity and become a unit of provisioning the environment. It would include issue and code tracking references, key performance indicators, x-rays and service map references, alerts and notifications and continuously updated with each deployment.

TCO of an IaC for a complex deployment does not include the person-hours required to keep it in a working condition and to assist with redeployments and syncing. One-off investigations are just too many to count on a hand in the case when deployments are large and complex. The sheer number of resources and their tracking via names and identifiers can be exhausting. A sophisticated CI/CD for managing accounts and deployments is good automation but also likely to be run by several contributors. When edits are allowed and common automation accounts are used, it can be difficult to know who made the change and why. All of these shortcomings can be overcome with a cloud IaC data model that is continuously updated via each pipeline-based deployment and encompasses silo'ed views of the numerous pipelines and repositories that exist while providing a base for canning repeated queries.

Some flexibility is required to make judicious use of automation and manual interventions to keep the deployments robust. Continuously updating the IaC and its knowledge base, especially by the younger members of the team, is not only a comfort but also a necessity. The more mindshare the IaC data model gets, the more likely it will reduce the costs associated with maintaining the IaC and dispel some of the limitations mentioned earlier.

As with all solutions, scope and boundaries apply. It is best not to let IaC, or its data model spread out so much that the high priority and severity deployments get affected. It can also be treated like any asset with its own index, model, documentation, and co-pilot.

IaC regulation and compliance

A simple anecdote might justify a discussion on this topic. When overseeing the deployment of many web applications and apis for an organization, I was using a common module for describing their infrastructure-as-code. This definition has a property called app_settings that the developers of the web applications and services use to save configuration and secrets that are read by their code. When we make changes to the infrastructure such as for consistency and conformance across these assets, the IaC compiler from the CI/CD pipeline would find the differences between the resources as they exist in the cloud and determine the changes to apply. Since the account used with the pipeline must have permissions to modify the resources, it can read these app settings and find changes to the secrets that the developers made. These differences in the secrets were logged and cited on the code changes via GitHub, which resulted in violations across the organization's security standards. There were no changes made to the infrastructure to disclose the secrets but every time a developer changed them, they would get logged and reported. The resolution was to include a directive to the compiler to ignore the changes to app settings section of the resource.

The public cloud comes with numerous purview techniques for code and data, but infrastructure regulation and compliance are seldom addressed. By its nature, the infrastructure must meet the standards together with those for the applications. For example, in the above case, the developers needed to be informed that app settings are a good choice for storing non-sensitive application settings, such as endpoint locations, sizing, flags, etc. but a key vault is a better choice for storing sensitive information, such as encryption keys, certificates, passwords, etc. They could be given both options, by creating app settings that reference secrets stored in key vault. In this way, they can maintain secrets apart from their app's configuration and access them like any other app setting or connection string in the code.

The governance requirements of infrastructure-as-code (IaC) can vary depending on the specific needs and context of an organization. However, here are some common governance considerations for IaC:

1. Version control: Implement a version control system to track changes made to infrastructure code, enabling teams to collaborate, review, and revert changes when needed.
2. Access controls: Define and enforce access controls to ensure that only authorized individuals can modify infrastructure code. This helps maintain security and prevent unauthorized changes.
3. Code reviews: Establish a code review process to ensure that infrastructure code adheres to best practices, standards, and security guidelines. Code reviews can help identify issues, ensure consistency, and improve the quality of the infrastructure code.
4. Testing and validation: Implement automated testing and validation processes to verify the correctness and reliability of infrastructure code changes. This can include unit testing, integration testing, and end-to-end testing to catch potential issues early on.
5. Documentation: Maintain comprehensive documentation for infrastructure code, including its purpose, dependencies, configurations, and any relevant operational considerations. This helps facilitate understanding, troubleshooting, and future modifications.
6. Change management: Implement a change management process to govern the introduction of infrastructure code changes into production environments. This includes proper planning, testing, and approvals to minimize the risk of disruption or downtime.
7. Compliance and auditing: Ensure that infrastructure code adheres to relevant regulatory and compliance requirements. This may involve periodic audits, security assessments, and documentation of compliance controls.

8. Monitoring and logging: Implement monitoring and logging mechanisms to track the behavior and performance of infrastructure deployments. This helps detect and troubleshoot issues, as well as provides valuable insights for optimization and capacity planning.
9. Continuous improvement: Foster a culture of continuous improvement by regularly reviewing and refining governance practices for IaC. This involves learning from past experiences, incorporating feedback, and adapting to evolving requirements and technologies.

It is important to note that these requirements may vary based on the size, complexity, and industry-specific regulations of an organization. It is recommended to consult with relevant stakeholders and experts to tailor governance requirements to specific needs.

Taking the example of Azure, code and data compliance and governance in Azure refers to the practices and features such as:

1. Compliance: Azure provides a range of compliance certifications and attestations, including ISO 27001, HIPAA, GDPR, and more. These certifications demonstrate Azure's commitment to meeting the highest standards of security and privacy.
2. Data Protection: Azure offers various tools and features to help protect your data. This includes encryption at rest and in transit, data masking, and data loss prevention (DLP) capabilities. Azure also provides Azure Key Vault, a secure key management service, to safeguard encryption keys.
3. Privacy: Azure helps organizations comply with privacy regulations by offering privacy controls, such as Azure Information Protection, which enables classification and protection of sensitive data. Azure also provides tools to manage data subject requests, consent, and data retention policies.
4. Governance: Azure provides governance capabilities to help organizations enforce policies and manage their resources

effectively. Azure Policy allows you to define and enforce rules and compliance standards across your Azure environment. Azure Blueprints enables the creation of reusable templates to ensure consistent deployment of resources.
5. Auditing and Monitoring: Azure offers auditing and monitoring capabilities to track and monitor activities within your Azure environment. Azure Monitor provides visibility into the performance and health of your resources, while Azure Security Center offers threat detection and security recommendations.
6. Data Residency: Azure allows you to choose the geographic region where your data is stored, helping you comply with data residency requirements and regulations specific to your industry or country.

Overall, Azure provides a comprehensive set of tools and features to help organizations achieve data compliance and governance, ensuring their data remains secure, private, and compliant with industry regulations.

Authorization

Infrastructure can be visualized in two layers for the purpose of Authorization – the control plane layer and the data plane layer.

The control plane layer is about context and access over the data and resources and primarily deals with organization, namespace and metadata. It secures and maps the users to membership directories via the Authorization component that can include an external Identity provider or an IAM solution. Many open-source infrastructures provide the abilities to delegate IAM access to others as a solution or externalize the membership directory to products such as Active Directory or Lightweight Directory Access Protocol. A tenant is used to refer to this dedicated and isolated instance of membership directory and associated IAM service and is provisioned as soon as a customer joins the public cloud with a subscription. Each tenant has its own identity and access management scope, and is distinct and separate from other tenants

The Data plane layer is all about data at rest and transit and who can do so. Often, this involves partitioning on the data layer with separate partitions, tables, columns, identifiers and labels on the data storage schema and topics such as Message Queues, Kafka, tags, domains and other data.

Authorization affects the whole resource and consequently requires its entities to be globally unique. Policies can include role-based access control with tenant management, relationship-based access control over a hierarchy where tenants become root-level relationship or attributes based access control with tenancy as an attribute which we will describe shortly.

In Azure public cloud, for example, a role is a set of permissions that define what actions a security principal (such as a user, group, or service principal) can perform within Azure resources. Roles are used in conjunction with Azure Role-Based Access Control (RBAC) to grant specific capabilities to users or groups at various levels of scope, such as a subscription, resource group, or individual resource.

While role-based access control is the default policy model sought widely, relationship-based access control can help to describe complex relationships between resources in a system. This approach facilitates a "policy-as-a-data" realm as opposed to "policy-as-a-code" realm. The policy is split between declarative rules and logic and can be visualized using a relationship graph and a hierarchical structure for assets such as folders and documents. Hierarchy can help with inheritance. For example,

```
package example.rebac.edit.__fileSystemResource

fileSystemResource = ds.object({
   "key": input.resource.fileSystemResource,
   "type": "file-system-resource",
})

user = ds.object({
   "key": input.user.id,
   "type": "user",
})

allowed {
   ds.check_permission({
      "sub_id": user.id,
      "obj_type": "file-system-resource",
      "permission": "can-edit",
      "obj_id": fileSystemResource.id
   })
}
```

Attribute-based access control emphasizes the use of attributes associated with the subject, the resource or the environment to determine whether access is granted or not. This results in a fine-grained authorization model and when these attributes have dynamic values, there is a lot of flexibility achieved. Databases implement row-level security, so this has some precedence. Authorization models can also be combined.

Multi-tenant software-as-a-service infrastructure deployments do not need to implement authorization from scratch. Often there is an existing library, product or solution that can be integrated into the Software-as-a-service. If it must be implemented, there are multiple options. Open Policy Agent (OPA) can be leveraged as an authorization microservice. Open Policy Administration Layer (OPAL) enables one to manage the authorization layer at scale, using PubSub topics and JSON data format. Solutions like Permit.IO provide a way to integrate seamlessly for tenants.

Privilege and permissions: A case study

This case study follows role-based access control from preceding sections. It uses a simple problem to articulate some of the concepts from RBAC. An Azure Machine Learning workspace administrator wants to create different connections to external data stores with the help of datastore objects that encapsulate the connection parameters and settings. But the administrator wants to secure these objects such that user A gets one datastore but not the others and user B gets another datastore but not anything else. Permissions are granted by roles and since built-in roles from Azure such as 'Owner', 'Contributor' and 'Reader' roles encompass all data store objects, both users A and B are granted custom role(s) where individual permissions can be managed. A sample permissions appears as: Microsoft.MachineLearningServices/workspaces/datastores/read

However, the administrator faces a problem that this permission does not say Datastore1/read or Datastore2/read. In fact, both users must get generic datastores/read permission that they cannot do without if they must have access to their respective datastores.

The solution to this problem is fairly simple. There are no datastores created by the administrator. Instead, the users create the datastores programmatically passing it either the self-describing credentials procured independently such as 'Shared-Access-Signature Token' to an external data storage or an account key that grants full access to the bearer. Either way, they must have access to their desired scope, say, storage-account/container/path/to/file and have a role specific to that data storage which gives privilege to look up such a credential at their choice of scope.

The creation and use of datastores are just like that of credentials or connection objects required for a database. As long as the users manage it themselves, they can reuse it at will.

If the administrator must be tasked with isolating access to the users to their workspace components and objects, then two workspaces will be created and assigned to groups to which these users can subscribe individually.

Some publicly available directions on this topic appear to falsely state that custom roles and Role-based Access Control will solve this for the administrator. Assertions like "By properly configuring RBAC, we can control access to datastores and other resources within the workspace" are misleading because the differentiation is being made to the objects of the same kind. There may be guidance on other mechanisms available to administrators that might include access control at external resource, generating and assigning different SAS tokens as secrets, generating virtual network service endpoints, exposing datastores with fine-grained common access, or using monitoring and alerts to detect and mitigate potential security threats. It is also possible to combine a few of the above techniques but isolation of user access is the simplest technique.

Troubleshooting of role assignments via IaC: a case study

One of the least expected but most frequent problems is the Http Status Unauthorized error code when making web requests from one resource to another. It refers to an action that was not authorized or an unreachable target for taking the action. The corresponding http status code is 401. When this error occurs in a complex system, the process of elimination alone takes a lot of time and effort. On the other hand, the same error can be helpful to building a system with the least privileges by constantly verifying that unintended access is indeed forbidden. Let us review a case study pertaining to restricted data access on a storage account.

The infrastructure-as-code, aka IaC, for deploying a storage account in commercial systems is usually not complete without the use of a corresponding role assignment and source restrictions in reaching the account over the public and private networks. When a storage account is deployed as a data lake with hierarchical file system, it usually has two names to reach it. For example, binary large objects or blobs for short can be accessed programmatically from the storage account by the account's name as https://storageaccount.blob.core.windows.net and the files can be addressed to https://storageaccount.dfs.core.windows.net. If the storage account must only be accessed by private network, then there must be a private endpoint specific to both DNS names. Checking that the source subnet or its public IP address are allow-listed on the storage account facilitates the ruling out of network as a potential culprit.

Then, role-based access control can be studied. In this case, the storage account has multiple containers, and each container has been assigned Access Control Lists with read-write-execute permissions to different groups. The idea behind this is to consolidate the containers in the same storage account for the sake of consistency enforcement in layouts per container from an infrastructure perspective but allow different teams to own their respective containers. The members of the groups allowed on to each container must also be granted a role of Reader to the storage account that lets them view the storage account on their management portal from the browser. When they navigate to it, the contents of only their container are visible to them and they can take actions to download and upload.

This might often be surprising to many that the Reader role being just a control plane built-in role having permissions only to list and read the account and having no data plane permissions to allow read and write of contents in the container, still permits the user to do that. This is not a defect but a feature. Roles and ACLs both grant the ability to read and write but while role allows a blanket permission across all the contents of the storage accounts, ACLs are scoped and indeed grant the ability to read and write. Should the role have blanket permissions at the data plane level, the ACLs are conveniently skipped.

The trouble arises when an allowed member of a group tries to download an item from their container in the storage account and gets a Forbidden error code. The programmatic way of doing this leverages a construct as for instance in the case of Azure public cloud, called DefaultAzureCredentials. With these credentials, when a client reads the blob or the file is instantiated using the Software Development Kit shipped with the Azure Storage Account, the call is already authenticated by virtue of the calling identity of the principal. However, the error forbidden comes from authorization of the read action.

Given the assignment to the Reader role and the ACL, authorization is granted but the forbidden error is misleading. There are usually two steps to resolving this. The first step involves verifying the identity at the client's end and the second involves inspecting the role and the ACL on the target. With the credentials acquired from the constructor, a method on the credentials object can be invoked to retrieve an access token from the authenticating endpoint https://management.azure.com. Then this token can be interpreted to view the claims. One of the claim types will be an email and this will indicate the caller. If the expected and the actual email match, then there is no error on the caller side otherwise the environment must be checked for proper configuration, so that the credentials object can be properly constructed.

At the target side, the check access functionality of the IAM feature management menu item can be leveraged to find out if the calling principal has at least the minimum role required to communicate with the target. Once this is verified, then the ACLs can be inspected to ensure that the action is authorized. Failure in the role match or the ACL check requires suitable remedy.

The logged-in identity is verified against the identity provider and passed through to the target. Roles and ACLs must authorize this identity prior to data plane operation. If the user tries actions not governed by identity such as generating a Shared Access Signature link for their container item that encapsulates an authentication and authorization segment in the link for use by the bearer, then such an action falls outside the integrated identity-based and role-based access control. By virtue of SAS URL being an alternative to integrated authentication and authorization, read and write permissions granted to SAS URL requires the associated principal doing so to have data plane roles and permissions. Granting those permissions disrespects the ACLs. While SAS URLs can be beneficial to those principals, who are typically data scientists, as they overcome the challenges of isolated networks, broken passthroughs and heterogenous data sources and destinations during the initial onboarding, mature systems often streamline identity-based access.

Automated Cloud IaC using Copilot

A Copilot is an AI companion that can communicate with a user over a prompt and a response. It can be used for various services such as Azure and Security, and it respects subscription filters. Copilots help users figure out workflows, queries, code and even the links to documentation. They can even obey commands such as changing the theme of a user interface to light or dark mode. Copilots are well-integrated with many connectors and types of data sources supported. They implement different Natural Language Processing models and are available in various flagship products such as Microsoft 365 and GitHub. They can help create emails, code, and collaboration artifacts faster and better.

Copilots can write IaC code just like any other language. It follows the same precedence as a GitHub Copilot that helps developers write code in programming languages. It is powered by the OpenAI Codex model, which is a modified production version of the Generative Pre-trained Transformer-3 aka (GPT-3). The GPT-3 AI model created by OpenAI features 175 billion parameters for language processing. This is a collaboration effort between OpenAI, Microsoft and GitHub.

A copilot can be developed with no code using Azure OpenAI studio. We just need to instantiate a studio, associate a model, add the data sources, and allow the model to train. The models differ in syntactic or semantic search. The latter uses a concept called embedding that discovers the latent meaning behind the occurrences of tokens in the given data. So, it is more inclusive than the former. A search for time will specifically search for that keyword in a lexical search but a search for clock will include references to time with a model that uses embeddings. Either way, a search service is required to create an index over the dataset because it facilitates fast retrieval. A database such as Azure Cosmos DB can be used to assist with vector search.

At present, all these resources are created in a cloud, but their functionality can also be recreated on a local Windows machine using the same components with the slight difference that they are generally offered as extensions via the Integrated Development Environment aka IDE where end-users write code. One does not need to resort to open source because public clouds go to extra lengths to make developer tools and environment convenient. This helps to train the model in documents that are available only locally. Usually, the time to set up the resources is only a couple of minutes but the time to train the model on all the data is the bulk of the duration after which the model can start making responses to the queries sent. The time for the model to respond once it is trained is usually in the order of a couple of seconds, if not less. A cloud storage account has the luxury to retain documents indefinitely and with no limit to size but the training of a model on the corresponding data accrues cost and increases with the size of the data ingested to form an index. Copilots can be fun addition to wherever chat like question and answers are required especially for customer touchpoints. The technology behind copilots is Generative AI and we will get to that next.

Emerging Trends and Generative AI

Generative Artificial Intelligence (AI) refers to a subset of AI algorithms and models that can generate new and original content, such as images, text, music, or even entire virtual worlds. Unlike other AI models that rely on pre-existing data to make predictions or classifications, generative AI models create new content based on patterns and information they have learned from training data.

One of the most well-known examples of generative AI is Generative Adversarial Networks (GANs). GANs consist of two neural networks: a generator and a discriminator. The generator network creates new content, while the discriminator network evaluates the content and provides feedback to the generator. Through an iterative process, both networks learn and improve their performance, resulting in the generation of more realistic and high-quality content.

Generative AI has made significant advancements in various domains. In the field of computer vision, generative models can create realistic images or even generate entirely new images based on certain prompts or conditions. In natural language processing, generative models can generate coherent and contextually relevant text, making them useful for tasks like text summarization, translation, or even creative writing.

However, it is important to note that generative AI models can sometimes produce biased or inappropriate content, as they learn from the data they are trained on, which may contain inherent biases. Ensuring ethical and responsible use of generative AI is an ongoing challenge in the field.

Generative AI also presents exciting opportunities for creative industries. Artists can use generative models as tools to inspire their work or create new forms of art. Musicians can leverage generative AI models to compose music or generate novel melodies.

Overall, generative AI holds great potential for innovation and creativity, but it also raises important ethical considerations that need to be addressed to ensure its responsible and beneficial use in various domains.

Some examples of text generation models include ChatGPT, Copilot, Gemini, and LLaMA which are often collectively referred to as chatbots. They generate human-like responses to queries. Image generation models include Stable Diffusion, Midjourney, and DALL-E which create images from textual descriptions. Video generation models include Sora which can produce videos based on prompts. Other domains where Generative AI finds applications are in software development, healthcare, finance, entertainment, customer service, sales, marketing, art, writing, fashion, and product design.

A Code Copilot is an AI-powered code completion tool developed by organizations like GitHub in collaboration with OpenAI. It is designed to assist developers in writing code by suggesting relevant snippets and completing code lines. Copilot uses machine learning models trained on a vast amount of publicly available code from various sources, including GitHub repositories.

Among the code copilots, GitHub Copilot stands out on several factors. This copilot generates context-aware code suggestions as you type, helping you write code faster and reducing the need to search for code examples or documentation. It support for multiple programming languages, including Python, JavaScript, TypeScript, Ruby, Go, and more. It adapts to the specific syntax and conventions of each language. It suggests code completions based on the code you have already written, making it easier to navigate complex APIs or remember function names and argument orders. It can be as fine as generating line-by-line code or entire blocks of code based on the provided context, reducing the time spent on repetitive coding tasks. It can also help with code formatting by suggesting improvements to the structure, indentation, and overall readability of your code. It integrates with popular code editors and Integrated Development Environments, such as Visual Studio Code, allowing developers to seamlessly incorporate it into their existing workflow. GitHub Copilot is available as a subscription service, and it requires an active GitHub account to use.

While the copilot is helpful and increases productivity by leaps and bounds, its responses vary in their quality and satisfaction to developers' prompts. This is often an iterative process and fortunately one that comes with plenty of tips for taking up the responses a notch in the form of prompt engineering. The pros and cons of the code pilot also varies with use cases but most would agree that they speed up the development process by suggesting snippets and completing lines of code, reducing the time spent on manual typing and searching for code examples. Since coding is often associated with samples and documentations that are publicly available, the copilot benefits the developers by suggesting the best practices and coding conventions, helping them write cleaner and more maintainable code. They can also serve as a valuable learning resource, especially for iterations by beginners or those less familiar with a particular programming language or framework. This cannot be understated as it has been a blessing for innovators and product managers to not rely on engineers to build prototypes and test them. Copilot can also provide insights into different coding patterns and approaches, catch syntax errors and offer corrections in real-time, minimizing the chances of introducing bugs. It is particularly helpful to explore new libraries, APIs, or frameworks by providing relevant code examples and suggestions, allowing developers to experiment with unfamiliar concepts more easily.

It is also important to understand the limitations of the copilot as some of unhealthy habits that form include relying too heavily on code copilots to the point that generated code is not fully understand. Many fail to review and verify the suggested code for correctness and security. If the person using the copilot is not clear or accurate in the prompts or has vague context or requirements of a project, the copilot can respond with incorrect suggestions. Developers should exercise caution and use their judgment to ensure the generated code aligns with their intentions.

By their nature, copilots rely on existing code patterns and examples, which can limit creativity and originality in coding. It is important to balance the use of code copilots with personal problem-solving and innovation. They often require an internet connection to function, which can be a limitation when working in offline or restricted environments. Finally, not all programming languages are supported or integrated with a specific development environment or code editor which may limit their usefulness in certain projects or setups. Overall, string the right balance between utilizing their suggestions and maintaining a deep understanding of the code being written, ultimately rests with the user and not the copilot.

Infrastructure-as-a-code is just another form of code where human written configuration files are used as training samples for the code copilot to generate responses based on the prompts. With the integration of the code pilot into the IDE and the language aware context available to the code copilot, it is now easy to get representative responses for good prompts. All the pros and cons of using the code pilot also applies to IaC and the infrastructure engineers must master prompt engineering as well as demonstrate heightened awareness of the strengths and limitations of the code copilot.

One advantage that works in the favor of code pilots is the granularity of the IaC code in terms of resource types and attributes. By their exhaustive nature of describing virtually all code types and attributes via documentation and samples, many of the responses from the copilot differ truly little from how an infrastructure engineer would write them.

However, unlike application code and object-oriented programming that deals with significant code reorganization and adherence to reusability and Do-not-repeat-yourself aka DRY principle, IaC is all about composability and repeatability. While this works in favor of cut-and-paste prompt responses from the code copilot, there is little training and even some tribal art on what constitutes a repeatable deployment stamp of infrastructure. At the higher-level constructs and layers of architecture, there are just not enough samples for IaC. As it is commonly known to many infrastructure provider languages, there are multiple stages of 'planning' and 'applying' IaC code and each involves a lot of vetting, trials and errors.

One way to overcome that is to build an in-house knowledge base of training samples that include previous code repositories and a custom copilot that can be fallback to the online web-based code copilots for consolidated responses. These "pseudo" resources representing complex but repeatable collections of resources, can be made first-class citizens of the training samples. With the analogy to text copilots, where jargons and topic words represent paragraphs and narratives about a discipline or focus area and leveraging the same co-occurrence based latent semantics that are captured via embeddings, the first-class entities and pseudo resources can be copied or put-together to form even large deployments.

There is a lot of road to cover before having a copilot be as effective as a veteran infrastructure engineer of an enterprise given the limited articulation of the resources and the size limits of the responses, but customizations can really fine-tune the copilot model because both the training samples and the model's hyper-parameters are in the hands of the infrastructure engineers.

Popularity of chatbots

Chatbots as a citizen of the public cloud has become such a ubiquitous commodity, it is widely adopted by folks from all disciplines including the consumers of infrastructure. For example, with the surge of data science and analytics projects, many data scientists are required to build a chatbot application for their data projects that are not at the infrastructure level where the computer, storage and networking are put into a workspace for these subscribers. Let us see some of the ways for the data scientists to do that as self-service. Let us say that they use a Databricks workspace, and the data is available via the catalog and delta lake and an all-purpose compute-cluster has been provisioned as dedicated to this effort. The example/tutorial to build a rudimentary chatbot comes from Databricks official documentation but it is put-together in the Azure public cloud where the querying capability and user-interface can be hosted outside the workspace and directly in the cloud.

Part 1.

The development of the model is best suited to the workspace. It proceeds this way:

Step 1. Set up the environment:

%pip install transformers sentence-transformers faiss-cpu

Step 2. Load the data into a Delta table:

```
from pyspark.sql import SparkSession
spark = SparkSession.builder.appName("Chatbot").getOrCreate()
# Load your data
data = [
```

```
{"id": 1, "text": "What is Databricks?"},
{"id": 2, "text": "How to create a Delta table?"}
]
df = spark.createDataFrame(data)
df.write.format("delta").save("/mnt/delta/chatbot_data")
```

Step 3. Generate embeddings using a pre-trained model:

```
from sentence_transformers import SentenceTransformer
model = SentenceTransformer('all-MiniLM-L6-v2')
texts = [row['text'] for row in data]
embeddings = model.encode(texts)
# Save embeddings
import numpy as np
np.save("/dbfs/mnt/delta/embeddings.npy", embeddings)
```

Step 4. Use FAISS to perform vector search over the embeddings.

```
import faiss
# Load embeddings
embeddings = np.load("/dbfs/mnt/delta/embeddings.npy")
# Create FAISS index
index = faiss.IndexFlatL2(embeddings.shape[1])
index.add(embeddings)
# Save the index
faiss.write_index(index, "/dbfs/mnt/delta/faiss_index")
```

Step 5. Create a function to handle user queries and return relevant responses.

```
def chatbot(query):
    query_embedding = model.encode([query])
    D, I = index.search(query_embedding, k=1)
    response_id = I[0][0]
    response_text = texts[response_id]
    return response_text
```

```
# Test the chatbot
print(chatbot("Tell me about Databricks"))
```

Step 6. Deploy the chatbot as

Option a) Databricks widget

```
dbutils.widgets.text("query", "", "Enter your query")
query = dbutils.widgets.get("query")

if query:
    response = chatbot(query)
    print(f"Response: {response}")
else:
    print("Please enter a query.")
```

Option b) a rest api
```
from flask import Flask, request, jsonify

app = Flask(__name__)

@app.route('/chatbot', methods=['POST'])
def chatbot_endpoint():
    query = request.json['query']
    response = chatbot(query)
    return jsonify({"response": response})

if __name__ == '__main__':
    app.run(host='0.0.0.0', port=5000)
```

Step 7. Test the API:

For option a) use the widgets to interact with the notebook:

```
# Display the widgets
```

```
dbutils.widgets.text("query", "", "Enter your query")
query = dbutils.widgets.get("query")
if query:
    response = chatbot(query)
    displayHTML(f"<h3>Response:</h3><p>{response}</p>")
else:
    displayHTML("<p>Please enter a query.</p>")
```

For option b) make a web request:

```
curl -X POST http://<your-databricks-url>:5000/chatbot -H "Content-Type: application/json" -d '{"query": "Tell me about Databricks"}'
```

Part 2.

This is about hosting the querying client and user interface and comprises the following:

```
import openai, os, requests
openai.api_type = "azure"
# Azure OpenAI on your own data is only supported by the 2023-08-01-preview API version
openai.api_version = "2023-08-01-preview"

# Azure OpenAI setup
openai.api_base = "https://azai-open-1.openai.azure.com/" # Add your endpoint here
openai.api_key = os.getenv("OPENAI_API_KEY") # Add your OpenAI API key here
deployment_id = "mdl-gpt-35-turbo" # Add your deployment ID here

# Azure AI Search setup
search_endpoint = "https://searchrgopenaisadocs.search.windows.net"; # Add your Azure AI Search endpoint here
```

```
search_key = os.getenv("SEARCH_KEY"); # Add your Azure AI Search
admin key here
search_index_name = "undefined"; # Add your Azure AI Search index
name here

def setup_byod(deployment_id: str) -> None:
    """Sets up the OpenAI Python SDK to use your own data for the
    chat endpoint.
    :param deployment_id: The deployment ID for the model to use with
    your own data.
    To remove this configuration, simply set openai.requestssession to
    None.
    """
    class BringYourOwnDataAdapter(requests.adapters.HTTPAdapter):
        def send(self, request, **kwargs):
            request.url = f"{openai.api_base}/openai/deployments/{deployment_id}/extensions/chat/completions?api-version={openai.api_version}"
            return super().send(request, **kwargs)
    session = requests.Session()

    # Mount a custom adapter which will use the extensions endpoint for
    any call using the given `deployment_id`
    session.mount(
        prefix=f"{openai.api_base}/openai/deployments/{deployment_id}",
        adapter=BringYourOwnDataAdapter()
    )
    openai.requestssession = session
setup_byod(deployment_id)

message_text = [{"role": "user", "content": "What are the differences between Chat GPT 3 and 4?"}]

completion = openai.ChatCompletion.create(
    messages=message_text,
```

```
    deployment_id=deployment_id,
    dataSources=[# camelCase is intentional, as this is the format the API expects
        {
            "type": "AzureCognitiveSearch",
            "parameters": {
                "endpoint": search_endpoint,
                "key": search_key,
                "indexName": search_index_name,
            }
        }
    ]
)
print(completion)
```

The user interface is simpler with code to host the app service as a Single Page WebApplication developed with React framework:

```
npm install @typebot.io/js @typebot.io/react
import {Standard} from "@typebot.io/react";

const App = () => {
  return (
    <Standard
      typebot="basic-chat-gpt-civ35om"
      style={{width: "100%", height: "600px"}}
    />
  );
};
```

This concludes the creation of a chatbot function using a workspace in the public cloud.

Managing copilots

Having touted the suitability of a copilot for every domain and discipline, this section of the series on copilots focuses on their proliferation and the internal processes required to manage them. As with any flight, a copilot is of assistance only to the captain responsible for the flight to be successful. If the captain does not know where she is going, then the copilot immense assistance will still not be enough. It is of secondary importance that the data that a copilot uses might be prone to bias or shortcomings and might even lead to so-called hallucinations for the copilot. Copilots are after all large language models that work entirely on treating data as vectors and leveraging classification, regression, and vector algebra to respond to queries. They don't build a knowledge graph and do not have the big picture on what business purpose they will be applied to. If the purpose is not managed properly, infrastructure engineers might find themselves maintaining many copilots for different use cases and even reducing the benefits of where a consolidated one would have sufficed for different personas.

Consolidation of large language models and their applications to different datasets is only the tip of the iceberg that these emerging technologies have provided as instruments for the data scientists. Machine Learning pipelines and applications are as diverse and silo'ed as the datasets that they operate on and they are not always present in data lakes or virtual warehouses. Consequently, a script or a prediction api written and hosted as an application does not make the best use of infrastructure for customer convenience in terms of interaction streamlining and more effective touch points. This is not to say that different models cannot be used or that the resources don't need to proliferate or that there are some cost savings with consolidation, it is about business justification of the number of copilots needed. When we work backwards from what the customer benefits or experiences, one of the salient features that works in favor of infrastructure management is that less is more.

Hyperconvergence of infrastructure for various business purposes when those initiatives are bought into by stakeholders that have both business and technical representations makes the investments more deliberate and fulfilling.

And cloud or infrastructure management is not restrictive to experimentation, just that it is arguing against the uncontrolled experimentation and placing the customers in a lab. As long as experimentation and initiatives can be specific in terms of duration, budget and outcomes, infrastructure management can go the extra mile of cleaning up, decommissioning, and even repurposing so that technical and business outcomes go hand in glove.

Processes are hard to establish when the technology is emerging and processes are also extremely difficult to enforce as new standards in the workplace. The failure of six sigma and the adoption of agile technologies are testament to that. However, best practices for copilot engineering are easier to align with cloud best practices and current software methodologies in most workplaces.

LLM-as-a-judge

There is a growing need for dynamic, dependable, and repeatable infrastructure as the scope of deployment expands from small footprint to cloud scale. With emerging technologies like Generative AI, the best practices for cloud deployment have not matured enough to create playbooks. Generative Artificial Intelligence (AI) refers to a subset of AI algorithms and models that can generate new and original content, such as images, text, music, or even entire virtual worlds. Unlike other AI models that rely on pre-existing data to make predictions or classifications, generative AI models create new content based on patterns and information they have learned from training data. Many organizations continue to face challenges to deploy these applications at production quality. The AI output must be accurate, governed and safe.

Data infrastructure trends that have become popular in the wake of Generative AI include data lakehouses which brings out the best of data lakes and data warehouses allowing for both storage and processing, vector databases for both storing and querying vectors, and the ecosystem for ETL, data pipelines and connectors facilitating input and output of data at scale and even supporting real-time ingestion. In terms of infrastructure for data engineering projects, customers usually get started on a roadmap that progressively builds a more mature data function. One of the approaches for drawing this roadmap that experts observe as repeated across deployment stamps involves building a data stack in distinct stages with a stack for every phase on this journey. While needs, level of sophistication, maturity of solutions, and budget determines the shape these stacks take, the four phases are more or less distinct and repeated across these endeavors. They are starters, growth, machine-learning and real-time. Customers begin with a starters stack where the essential function is to collect the data and often involve implementing a drain. A unified data layer in this stage significantly reduces engineering bottlenecks. A second stage is the growth stack which solves the problem of proliferation of data destinations and independent silos by centralizing data into a warehouse which also becomes a single source of truth for analytics. When this matures, customers want to move beyond historical analytics and into predictive analytics. At this stage, a data lake and machine learning toolset come handy to leverage unstructured data and mitigate problems proactively. The next and final frontier to address is the one that overcomes a challenge in this current stack which is that it is impossible to deliver personalized experiences in real-time.

Even though it is a shifting landscape, the AI models are largely language models and some serve as the foundation for layers for increasingly complex techniques and purpose. Foundation models commonly refer to large language models that have been trained over extensive datasets to be generally good at some task(chat, instruction following, code generation, etc.) and they largely follow in two categories: proprietary (such as Phi, GPT-3.5 and Gemini) and open source (such as Llama2-70B and DBRX). DBRX for its popularity with Databricks platform that is ubiquitously found on different public clouds are transformer-based decoder large language models that are trained using next-token prediction. There are benchmarks available to evaluate foundational models.

Many end-to-end LLM training pipeline are becoming more compute-efficient. This efficiency is the result of a number of improvements including better architecture, network changes, better optimizations, better tokenization and last but not the least – better pre-training data which has a substantial impact on model quality.

A fine-grained mixture-of-experts (MoE) architecture typically works better than any single model. Inference efficiency and model quality are typically in tension: bigger models typically reach higher quality, but smaller models are more efficient for inference. Using MoE architecture makes it possible to attain better tradeoffs between model quality and inference efficiency than dense models typically achieve.

Companies in the foundational stages of adopting generative AI technology often lack a clear strategy, use cases, and access to data scientists. To start, companies can use off-the-shelf Learning Logistic Models (LLMs) to experiment with AI tools and workflows. This allows employees to craft specialized prompts and workflows, helping leaders understand their strengths and weaknesses. LLMs can also be used as a judge to evaluate responses in practical applications, such as sifting through product reviews.

Large Language Models (LLMs) have the potential to significantly improve organizations' workforce and customer experiences. By addressing tasks that currently occupy 60%-70% of employees' time, LLMs can significantly reduce the time spent on background research, data analysis, and document writing. Additionally, these technologies can significantly reduce the time for new workers to achieve full productivity. However, organizations must first rethink the management of unstructured information assets and mitigate issues of bias and accuracy. This is why many organizations are focusing on internal applications, where a limited scope provides opportunities for better information access and human oversight. These applications, aligned with core capabilities already within the organization, have the potential to deliver real and immediate value while LLMs and their supporting technologies continue to evolve and mature. Examples of applications include automated analysis of product reviews, inventory management, education, financial services, travel and hospitality, healthcare and life sciences, insurance, technology and manufacturing, and media and entertainment.

The use of structured data in GenAI applications can enhance their quality such as in the case of a travel planning chatbot. Such an application would use a vector search and feature-and-function serving building blocks to serve personalized user preferences and budget and hotel information often involving agents for programmatic access to external data sources. To access data and functions as real-time endpoints, federated and universal access control could be used. Models can be exposed as Python functions to compute features on-demand. Such functions can be registered with a catalog for access control and encoded in a directed acyclic graph to compute and serve features as a REST endpoint.

To serve structured data to real-time AI applications, precomputed data needs to be deployed to operational databases, such as DynamoDB and Cosmos DB as in the case of AWS and Azure public clouds respectively. Synchronization of precomputed features to a low-latency data format is required. Fine-tuning a foundation model allows for more deeply personalized models, requiring an underlying architecture to ensure secure and accurate data access.

Most organizations do well with an Intelligence Platform that helps with model fine-tuning, registration for access control, secure and efficient data sharing across different platforms, clouds and regions for faster distribution worldwide, and optimized LLM serving for improved performance. The choice of such Intelligence platforms should be such that it is simple infrastructure for fine-tuning models, ensuring traceability from models to datasets, and enabling faster throughput and latency improvements compared to traditional LLM serving methods.

Software-as-a-service LLMs aka SaaS LLMs are way more costly than those developed and hosted using foundational models in workspaces either on-premises or in the cloud because they need to address all the use cases including a general chatbot. The generality incurs cost. For a more specific use case, a much smaller prompt suffices and it can also be fine-tuned by baking the instructions and expected structure into the model itself. Inference costs can also rise with the number of input and output tokens and in the case of SaaS services, they are charged per token. Specific use case models can even be implemented with 2 engineers in 1 month with a few thousand dollars of compute for training and experimentation and tested by 4 human evaluators and an initial set of evaluation examples.

SaaS LLMs could be a matter of convenience. Developing a model from scratch often involves significant commitment both in terms of data and computational resources such as pre-training. Unlike fine-tuning, pre-training is a process of training a language model on a large corpus of data without using any prior knowledge or weights from an existing model. This scenario makes sense when the data is quite different from what off-the-shelf LLMs are trained on or where the domain is rather specialized when compared to everyday language or there must be full control over training data in terms of security, privacy, fit and finish for the model's foundational knowledge base or when there are business justifications to avoid available LLMs altogether.

Organizations must plan for the significant commitment and sophisticated tooling required for this. Libraries like PyTorch FSDP and Deepspeed are required for their distributed training capabilities when pretraining an LLM from scratch. Large-scale data preprocessing is required and involves distributed frameworks and infrastructure that can handle scale in data engineering. Training of an LLM cannot commence without a set of optimal hyperparameters. Since training involves high costs from long-running GPU jobs, resource utilization must be maximized. Even the length of time for training might be quite large which makes GPU failures more likely than normal load. Close monitoring of the training process is essential. Saving model checkpoints regularly and evaluating validation sets acts as safeguards.

Constant evaluation and monitoring of deployed large language models and generative AI applications are important because both the data and the environment might vary. There can be shifts in performance, accuracy or even the emergence of biases. Continuous monitoring helps with early detection and prompt responses, which in turn makes the models' outputs relevant, appropriate and effective. Benchmarks help to evaluate models but the variations in results can be large. This stems from a lack of ground truth. For example, it is difficult to evaluate summarization models based on traditional NLP metrics such as BLEU, ROUGE etc because summaries generated might have completely different words or word ordes. Comprehensive evaluation standards are elusive for LLMs and reliance on human judgment can be costly and time-consuming. The novel trend of "LLMs as a judge" still leaves unanswered questions about reflecting human preferences in terms of correctness, readability and comprehensiveness of the answers, reliability and reusability on different metrics, use of different grading scales by different frameworks and the applicability of the same evaluation metric across diverse use cases.

Finally, the system must be simplified for use with model serving to manage, govern and access models via unified endpoints to handle specific LLM requests.

Since chatbots are common applications of LLM, an example of evaluating a chatbot now follows. The underlying principle in a chatbot is Retrieval Augmented Generation and it is quickly becoming the industry standard for developing chatbots. As with all LLM and AI models, it is only as effective as the data which in this case is the vector store aka knowledge base. The LLM could be newer GPT3.5 or GPT4 to reduce hallucinations, maintain up-to-date information, and leverage domain-specific knowledge. Evaluating the quality of chatbot responses must take into account both the knowledge base and the model involved. LLM-as-a-judge fits this bill for automated evaluation but as noted earlier, it may not be at par with human grading, might require several auto-evaluation samples and may have different responsiveness to different chatbot prompts. Slight variations in the prompt or problem can drastically affect its performance.

RAG-based chatbots can be evaluated by LLM-as-a-judge to agree on human grading on over 80% of judgements if the following can be maintained: using a 1-5 grading scale, use GPT-3.5 to save costs and when you have one grading example per score and use GPT-4 as an LLM judge when you have no examples to understand grading rules.

The initial evaluation dataset can be formed from say 100 chatbot prompts and context from the domain in terms of (chunks of) documents that are relevant to the question based on say F-score. Using the evaluation dataset, different language models can be used to generate answers and stored in question-context-answers pairs in a dataset called "answer sheets". Then given the answer sheets, various LLMs can be used to generate grades and reasoning for grades. Each grade can be a composite score with weighted contributions for correctness (mostly), comprehensiveness and readability in equal proportions of the remaining weight. A good choice of hyperparameters is equally applicable to LLM-as-a-judge and this could include low temperature of say 0.1 to ensure reproducibility, single-answer grading instead of pairwise comparison, chain of thoughts to let the LLM reason about the grading process before giving the final score and examples in grading for each score value on each of the three factors. Factors that are difficult to measure quantitatively include helpfulness, depth, creativity etc. Emitting the metrics about correctness, comprehensiveness, and readability provides justification that becomes valuable. Whether we use a GPT-4, GPT-3.5 or human judgement, the composite scores can be used to tell results apart quantitatively. The overall workflow for the creation of LLM-as-a-judge is also similar to the data preparation, indexing relevant data, information retrieval and response generation for the chatbots themselves.

Data Infrastructure for Generative AI

Data infrastructure trends that have become popular in the wake of Generative AI include data lakehouses which brings out the best of data lakes and data warehouses allowing for both storage and processing, vector databases for both storing and querying vectors, and the ecosystem for ETL, data pipelines and connectors facilitating input and output of data at scale and even supporting real-time ingestion. In terms of infrastructure for data engineering projects, customers usually get started on a roadmap that progressively builds a more mature data function. One of the approaches for drawing this roadmap that experts observe as repeated across deployment stamps involves building a data stack in distinct stages with a stack for every phase on this journey. While needs, level of sophistication, maturity of solutions, and budget determines the shape these stacks take, the four phases are more or less distinct and repeated across these endeavors. They are starters, growth, machine-learning and real-time. Customers begin with a starters stack where the essential function is to collect the data and often involve implementing a drain. A unified data layer in this stage significantly reduces engineering bottlenecks. A second stage is the growth stack which solves the problem of proliferation of data destinations and independent silos by centralizing data into a warehouse which also becomes a single source of truth for analytics. When this matures, customers want to move beyond historical analytics and into predictive analytics. At this stage, a data lake and machine learning toolset come handy to leverage unstructured data and mitigate problems proactively. The next and final frontier to address is the one that overcomes a challenge in this current stack which is that it is impossible to deliver personalized experiences in real-time.

Even though it is a shifting landscape, the AI models are largely language models and some serve as the foundation for layers for increasingly complex techniques and purpose. Foundation models commonly refer to large language models that have been trained over extensive datasets to be generally good at some task(chat, instruction following, code generation, etc.) and they largely follow in two categories: proprietary (such as Phi, GPT-3.5 and Gemini) and open source (such as Llama2-70B and DBRX). DBRX for its popularity with Databricks platform that is ubiquitously found on different public clouds are transformer-based decoder large language models that are trained using next-token prediction. There are benchmarks available to evaluate foundational models.

Many end-to-end LLM training pipeline are becoming more compute-efficient. This efficiency is the result of a number of improvements including better architecture, network changes, better optimizations, better tokenization and last but not the least – better pre-training data which has a substantial impact on model quality.

A fine-grained mixture-of-experts (MoE) architecture typically works better than any single model. Inference efficiency and model quality are typically in tension: bigger models typically reach higher quality, but smaller models are more efficient for inference. Using MoE architecture makes it possible to attain better tradeoffs between model quality and inference efficiency than dense models typically achieve.

Companies in the foundational stages of adopting generative AI technology often lack a clear strategy, use cases, and access to data scientists. To start, companies can use off-the-shelf Learning Logistic Models (LLMs) to experiment with AI tools and workflows. This allows employees to craft specialized prompts and workflows, helping leaders understand their strengths and weaknesses. LLMs can also be used as a judge to evaluate responses in practical applications, such as sifting through product reviews.

Large Language Models (LLMs) have the potential to significantly improve organizations' workforce and customer experiences. By addressing tasks that currently occupy 60%-70% of employees' time, LLMs can significantly reduce the time spent on background research, data analysis, and document writing. Additionally, these technologies can significantly reduce the time for new workers to achieve full productivity. However, organizations must first rethink the management of unstructured information assets and mitigate issues of bias and accuracy. This is why many organizations are focusing on internal applications, where a limited scope provides opportunities for better information access and human oversight. These applications, aligned with core capabilities already within the organization, have the potential to deliver real and immediate value while LLMs and their supporting technologies continue to evolve and mature. Examples of applications include automated analysis of product reviews, inventory management, education, financial services, travel and hospitality, healthcare and life sciences, insurance, technology and manufacturing, and media and entertainment.

The use of structured data in GenAI applications can enhance their quality such as in the case of a travel planning chatbot. Such an application would use a vector search and feature-and-function serving building blocks to serve personalized user preferences and budget and hotel information often involving agents for programmatic access to external data sources. To access data and functions as real-time endpoints, federated and universal access control could be used. Models can be exposed as Python functions to compute features on-demand. Such functions can be registered with a catalog for access control and encoded in a directed acyclic graph to compute and serve features as a REST endpoint.

To serve structured data to real-time AI applications, precomputed data needs to be deployed to operational databases, such as DynamoDB and Cosmos DB as in the case of AWS and Azure public clouds respectively. Synchronization of precomputed features to a low-latency data format is required. Fine-tuning a foundation model allows for more deeply personalized models, requiring an underlying architecture to ensure secure and accurate data access.

Most organizations do well with an Intelligence Platform that helps with model fine-tuning, registration for access control, secure and efficient data sharing across different platforms, clouds and regions for faster distribution worldwide, and optimized LLM serving for improved performance. The choice of such Intelligence platforms should be such that it is simple infrastructure for fine-tuning models, ensuring traceability from models to datasets, and enabling faster throughput and latency improvements compared to traditional LLM serving methods.

Software-as-a-service LLMs aka SaaS LLMs are way more costly than those developed and hosted using foundational models in workspaces either on-premises or in the cloud because they need to address all the use cases including a general chatbot. The generality incurs cost. For a more specific use case, a much smaller prompt suffices and it can also be fine-tuned by baking the instructions and expected structure into the model itself. Inference costs can also rise with the number of input and output tokens and in the case of SaaS services, they are charged per token. Specific use case models can even be implemented with 2 engineers in 1 month with a few thousand dollars of compute for training and experimentation and tested by 4 human evaluators and an initial set of evaluation examples.

Networking: a case study

This section describes a case study with azure private dns zones and virtual network links for the purpose of demonstrating that that no single advice works for all usages. Virtual networks allow isolation between deployments as the address space is closed and has no relevance to those of other virtual networks whether overlapping or not. Typically, each virtual network is confined to a region and is zonal in its nature much like virtual machines and their scale sets.

Azure Private dns zones are global and are not confined to a region. The purpose of a private dns zone is to allow automatic name resolution to private IP addresses. For example, an entry for an AppService, say web-app-1, in the privatelink.azurewebsites.net allows that web-app-1 to be reached on its private IP address by virtue of a private endpoint even for a fully qualified domain name like web-app-1.azurewebsites.net which usually would have been resolved to its public IP address. It does this with the help of recordsets and virtual network links, two features that serve different purposes. The recordset allows the hostname to IP address conversion. The virtual network link allows for the name translation by prefixing with privatelink from callers in these associated virtual networks.

And that is where the difference lies between the number of private dns zones one must have. The answer depends on whether the callers to a particular resource will originate from as well as how many private IP addresses the resource will have. Typically, one address is sufficient within a virtual network to route the traffic to the resource. Therefore, if a resource has multiple private endpoints, one each in different virtual networks, it receives multiple private addresses. Each of these addresses can be associated with the hostname for a name resolution only once within a private dns zones as all address records are unique by hostname in the record sets of that zone. If there are multiple private IP addresses, then there will be as many dns zones for each of the records. Typically, callers will need only one name or IP address for the resource and these callers can all come through the same virtual network.

If we have a resource in a virtual network and we commission a similar resource in another virtual network before decommissioning the original one, then we could have the following choices:

A. A single global private dns zone with virtual links to old and new virtual networks
B. An existing and a new private dns zone with same name and in different resource groups but with virtual network links to existing and new virtual networks independently.

The advantage of A is that all the erstwhile callers from the old virtual network can resolve the private IP address of the new resource while B cannot. The advantage of B is that there is clear isolation and A can be decommissioned independent of the operations of the new virtual network, zone, and resource.

However, there is a twist. A virtual network link between one virtual network and similarly named private dns zones in separate resource groups can only be created once. This means that regardless of whether the virtual network link is created in an existing private dns zone by the same name as that of a new private dns zone, it will only be in one place.

That is why neither choice A nor B are mutually exclusive, and one must leverage the status quo before deciding to consolidate or separate the virtual network links in dns zones.

Understanding Workloads for business continuity and disaster recovery (aka BCDR)

The Azure public cloud provides native capabilities in the cloud for the purposes of business continuity and disaster recovery, some of which are built into the features of the resource types used for the workload. Aside from features within the resource type to reduce RTO/RPO (for a discussion on terms used throughout the BCDR literature) please use the references), there are dedicated resources such as Azure Backup, Azure Site Recovery and various data migration services such as Azure Data Factory and Azure Database Migration Services that provided a wizard for configuring the BCDR policies which are usually specified in a file-and-forget way. Finally, there are customizations possible outside of those available from the features of the resource types and BCDR resources which can be maintained by Azure DevOps.

Organizations may find that they can be more efficient and cost-effective by taking a coarser approach at a deployment stamp level higher than the native cloud resource level and one that is tailored to their workload. This section explores some of those scenarios and the BCDR solutions that best serve them.

Workload 1: Applications

Microservice framework form of deployment is preferred when the workload wants to update various services hosted as api/ui independently from others for their lifetime. Usually, there are many web applications, and a resource is dedicated to each of them in the form of an app service or a container framework. The code is either deployed via a pipeline directly as source code or published to an image that the resource pulls. One of the most important aspects peculiar to this workload is the dependencies between various applications. When a disaster strikes the entire deployment, they will not all work together even when restored individually in a different region without reestablishing these links. Take for example, the private endpoints that provide connectivity privately between caller-callee pairs of these services. Sometimes the callee is external to the network and even subscription and usually endpoint establishing the connectivity is manually registered. There is no single button or pipeline that can recreate the deployment stamp and certainly none that can replace the manual approval required to commission the private link. Since individual app services maintain their distinctive dependencies and fulfilment of functionality but cannot work without the whole set of app services, it is important to make them exportable and importable via Infrastructure-as-code aka IaC that takes into account parameters such as subscription, resource groups, virtual network, prefixes and suffixes in naming convention and recreates a stamp.

The second characteristic of this workload is that typically it will involve a diverse set of dependencies and stacks to host the various web applications that it does. There will not be any consistency, so the dependencies could range from a MySQL database server to producing and consuming jobs on a Databricks analytical workspace or an airflow automation. Consequently, the dependencies must be part of the BCDR story. Since this usually involves data and scripts, they should be migrated to the new instance. Migration and renaming are two pervasive activities for the BCDR of this workload type. Scripts that are registered in a source code repository like GitHub must be pulled and spun into an on-demand triggered or scheduled workflow.

Lastly, data used by these resources are usually proprietary and territorial in terms of ownership. This implies that the backup and restore of data might have to exist independently and as per the consensus with the owner and the users. MySQL data can be transferred to and from another instance via the Azure Database Migration Service so as to avoid the use of mysqldump command line with credentials or via GitOps via the az command to the database server instance with implicit login. An approach that suits the owner and users can be implemented outside the IaC.

One of the aspects that is not often called out is that these app services must be protected by web application firewall that conforms to the OWASP specifications. This is addressed with the use of an application gateway and FrontDoor. With slight differences between the two, both can be leveraged to switch traffic to an alternate deployment but only one of them is preferred to switch to a different region. FrontDoor has the capability to register a unique domain per backend pool member so that the application receives all traffic addressed to the domain at the root "/" path as if it were sent to it directly. It also comes with the ability to switch regions such as between centralus and east us 2. Application gateway, on the other hand, is pretty much regional with one instance per region. Both can be confined to a region by directing all traffic between their frontend and backends to go through the same virtual network. Networking infrastructure is probably the biggest investment that needs to be made up front for BCDR planning because each virtual network is specific to a region. Having the network up and running allows resources to be created on-demand so that the entire deployment for another region can be created only on-demand. As such an Azure Application Gateway or Front Door must be considered a part of the workload along with the other app services and planned for migration.

Workload #2: Hosting

An Azure Kubernetes instance that works more for rehosting on-premises apps and services than for the restructuring that the workload #1 serves. In this case, there is more consolidation and also significant encumbrance on now so-called "traditional" way of hosting applications and the logic that had become part of the kube-api server and data and state that was saved to persistent volume claims must now become part of the BCDR protection. A sample architecture that serves this workload can be referred to with the diagram below:

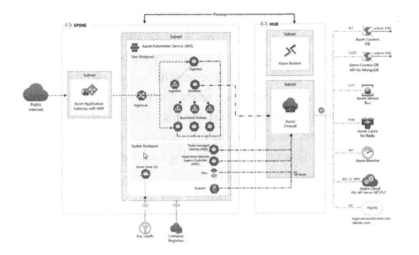

In this case, the BCDR can follow the pattern for AKS called out in the best practice patterns specific to this resource type.

Workload #3: Analytical workspaces:

As with most data science efforts, there will be some that require interactive deployment versus those that can be scheduled to run non-interactively. Examples of these workspaces include Azure Databricks and Azure Machine Learning. The characteristics of this kind of workload are that they are veritable ecosystems by themselves and one that relies heavily on compute and externalizes storage. Many workspaces will come with external storage accounts, databases, and Snowflake warehouses. Another characteristic is that these resources often require both public and private plane connectivity, so workspaces that are created in another region must re-establish connectivity to all dependencies, including but not limited to private and public source depot, container image repositories, and external databases and warehouses and by virtue of these dependencies being in different virtual networks, new private endpoints from those virtual networks become necessary. Just like AKS, the previous workload discussed above, manifesting all these dependencies that have been accrued over time might be difficult when they are not captured in IaC.

More importantly, it is the diverse set of artifacts that the workspace makes use of in terms of experiments, models, jobs, and pipelines which may live as objects in a catalog of the workspace but import and export of those objects to another workspace might not pan out the way as IaC does. With a diverse and distinct set of notebooks from different users and their associated dependencies, the task of listing these itself might be hard much less the migration to a new region. Users can only be encouraged to leverage the Unity Catalog and the version control of all artifacts external to the workspace, but they lack the rigor of databases. That said, spinning up a new workspace and re-connecting different data stores might provide a way for the users to be selective in what they bring to the new workspace.

Workload #4: Traffic

One of the goals in restoring a deployment after a regional outage is to reduce the number of steps in the playbook for enabling business critical applications to run. Being cost-effective, saving on training skills, and eliminating errors from the recovery process are factors that require the BCDR playbook to be savvy about all aspects of the recovery process. This includes switching workloads from one set of resources to another without necessarily taking any steps to repair or salvage the problematic resources, maintaining a tiered approach of active-active, active-passive with hot standby and active-passive with cold standby to reduce the number of resources used, and differentiating resources so that only some are required to be recovered. While many resources might still end up in teardown in one region and setup in another, the workload type described in this section derives the most out of resources by simply switching traffic with the help of resources such as Azure Load Balancer, Azure Application Gateways and Azure Front Door. Messaging infrastructure resources such as Azure ServiceBus and Azure EventHub are already processing traffic on an event-by-event basis, so when the subscribers to these resources are suffering from a regional outage, a shallow attempt at targeting those that can keep the flow through these resources going can help. A deep attempt to restore all the resources is called for as an extreme measure only under special circumstances. This way, there is optimum use of time and effort in the recovery.

Workload #5: Data stores:

This workload type is specific to the vast and diverse deployments for data at rest and in transit in the form of data stores, messaging publishers, subscribers, and pipelines, each with its own quirks and exacting procedures. The characteristic of this workload is that it is sticky just like the data that it stores. Over time, the inertia builds as the data and the resources housing and facilitating transfers grows. BCDR for such strategy is not always easy as a one-time cutover unless the organization is willing to spend on an active-active continuous replication system. Another characteristic is that out-of-band backup and recovery on ad hoc and scheduled basis require maintenance in the form of robustness against interruptions during lengthy durations. That said, data stores of various sizes can be categorized in T-shirt sizes of less than 1 GB, 10GB, 1TB and so on, with somewhat similar logic for handling data transfer. Public cloud provides the ability to create robust pipelines and often involves architecture that involves setting up an agent and a controller for the data and control plane operations, respectively. This meets another challenge in that the data is often integrated with on-premises stacks and solutions that require the agent to run on-premises. An Azure self-hosted integration runtime serves an example of this architecture and is fast and reliable for all kinds of access between the on-premises and the Azure public cloud. The trade-off is that dedicating integration runtimes and pipelines for the purpose of replication goes against the fast and easy recovery runbook steps, so cloud architects often consider centralized and governance-friendly data storage solutions that has in-built features for disaster recovery and cross-region replication. As mentioned earlier, such a deployment stamp might involve a variety of products and several of them, so some scripting and runbooks might require to be authored for the purposes of business continuity and disaster recovery. When applications need to change the name of the datastore or the connection string in their application code, it complicates the BCDR runbook. One way to resolve this is to use C-Name internal aliases so that they can be changed to point to the recovery instance with little or no effort.

Ownership

When deployments become complex, the maintenance of their IaC calls for human resource allocations. While the are many factors that are significant to the planning of this allocation, one of the characteristics of the deployments is that there is repeated copy-and-paste involved across different deployment stamps. When a person is allowed to focus on one set of resources within a stamp in a subscription, there will be very little changes required to be made that avoid inadvertent errors made from copy and paste across subscriptions. Every resource is named with a convention. A complex deployment increases the number of edits made across resources. When there is name variations introduced by the lines of business to differentiate deployment stamps and resources, even a modification of a resources across the subscription channels involves more copying than in the case when everything is self-contained within a subscription.

Another characteristic is that public cloud resource types require in-depth knowledge of how they work and some of them have sophisticated feature sets that it takes a while before a definition in the IaC for that resource type becomes the norm for deployment. It is in this regard that cloud engineer expertise in certain resource types become a sought-after skill for many teams and a convenience for the infrastructure management team to consolidate and direct questions and support requests to the same group of individuals. Usually, two people can act as primary and secondary owners of these resource types. When the resource type is complex such as the use of analytics workspaces that come with their compute and storage ecosystem, designating pairs of individuals, if not more, can help with bringing industry and community perspectives to the team via trainings and conferences.

A third characteristic of working the public cloud deployments with IaC from the management's point of view is the creation of active directory groups for individuals dedicated to working in owner, contributor and reader modes on deployments stamps and enabling the groups to be universal rather than global. The difference between groups created in these two modes is that one permits multi-domain environment access and changes to the membership trigger forest-wide replication which is helpful to ensure that permissions remain consistent across the forest. On-premises environments have traditionally used global groups since they are domain specific but with the migration to cloud resources, universal groups hold more appeal.

Securing access to resources via Active Directory groups also helps with the propagation of permissions and the ease of one-time registration to membership by individuals. When they leave, the access is automatically removed everywhere by the removal of membership and while this has remained true for most workplaces, it is even more pertinent when groups tend to be many for different purposes and creating well-known groups whose scope is tightly coupled to the resources they secure, help with less maintenance activities as individuals become empowered as needed to control the deployments of resources to the cloud.

A note about choices between public clouds

The examples throughout this book leverage resources in the Azure public cloud but cloud computing concepts are more or less similar across public clouds. Organizations leveraging multiple cloud solutions will do well to be true to the vendor-independent infrastructure concepts for the most part and with considerations for the customizations needed for a specific cloud. As an example, this section describes similar features on the Amazon Web Services public cloud which has enabled many organizations to move from on-premises datacenters to public cloud computing. They do away with their maintenance and total cost of ownership aka TCO and pay only for what they use, which also happens to be elastic, highly available and performant on a cloud scale. These cloud services are not just adaptable but also highly available all over the world. In addition, the cloud serves all kinds of audiences with its interfaces that include the management portal, command-line interfaces, REST based APIs and software development kit. Services like Identity Access Management aka IAM ensure that only the appropriate people in the company access just the resources they need. Services like S3 allow ubiquitous and versatile use cases with storage of data with many features to protect and govern their data. Services like Virtual Private Cloud enable organizations to isolate their infrastructure from others. To sum it up, AWS is cost-effective and highly flexible.

Cloud computing offers on-demand computing, storage, databases, and networking services through the internet. It is a cost-effective alternative to traditional data centers, as it is delivered on demand and requires only what is requested and used. Cloud storage allows for safe and secure preservation of data on remote servers, reducing the risk of data loss. For enormous amounts of structured data, a database is needed. Networking services are also necessary to connect these services. Amazon

Web Services (AWS) are primarily used through the AWS Management Console, command-line interface (CLI), or software development tool (SDK). To get started with AWS, users need to create an account and become a user. There are three ways to engage with AWS: the AWS Console, which is a web interface for managing resources, the CLI, which allows for automation of tasks, and the SDK, which supports various programming languages. Identity and Access Management (IAM) ensures that only the appropriate people in a company access the cloud services they need.

IAM (Information Access Management) is a cloud computing system that focuses on three core features: identification, authorization, and authentication. Identification involves identifying each user by providing a unique username and password, while authorization verifies the correct information provided. Amazon Simple Storage Service (S3) is a central offer that allows users to store and access data from anywhere. To store data in S3, users create an AWS bucket, which serves as a storage vehicle and can be accessed via the Management Console, CLI, or SDK. Objects in S3 can have a maximum size of 5 terabytes. AWS offers different storage classes based on data frequency, such as "S3 Standard," "S3 Intelligent-Tiering," "Infrequent Access," and "One Zone-Infrequent Access." A Virtual Private Cloud (VPC) is a private virtual space that is exclusive to the AWS account and is available in all regions. VPCs are private networks that can be accessed by the user and are available in all regions.

AWS's Elastic Cloud Compute (EC2) allows for the scaling of virtual machines (VMs) to meet specific computing needs, reducing physical infrastructure costs. EC2 provides "Infrastructure as a Service (IaaS)," where users rent virtual machines from AWS and configure their security arrangements to their specifications. AWS also offers an elastic load balancer (ELB) and autoscaling, which helps distribute incoming traffic in a balanced, manageable way. AWS Autoscaling automatically adds or removes capacity to optimize performance and keep costs low. There are various autoscaling approaches and policies, including manual adjustments, dynamic scaling, and predictive scaling.

Data comes in three types: structured, unstructured, and semi-structured. Structured data has a predetermined type and format, while unstructured data lacks a preset model. Semi-structured data can be changed without needing to alter all others. Storing small data can be managed through Excel sheets and tables, but substantial amounts of data require databases. AWS provides a variety of autoscaling approaches and policies to suit different business needs.

Databases come in several types, with relational databases ideal for structured data and non-relational databases for unstructured and semi-structured data. Organizations can choose between on-premises or cloud management, with cloud-based services like Amazon's Relational Database Service (RDS) being more efficient and cost-effective. Cloud security is crucial as it is a target for cyber-criminals. AWS provides secure cloud computing environments and 24/7 security measures, while users are responsible for their own security. AWS monitors data centers, provides cyberattack detection systems, and enforces rigorous employee authentication. Users must secure their operating systems and application data, regulate access, and comply with international security compliance issues.

Just-in-time access

Just-in-time (JIT) access, also known as just-in-time privileged access management (JIT PAM) is a security model used in say, Azure public cloud to grant temporary permissions to users for performing privileged tasks. This approach helps minimize the risk of unauthorized access by ensuring that elevated permissions are only available when needed and for a limited time. Users receive elevated permissions only for the duration necessary to complete specific tasks. Once the time expires, the permissions are revoked automatically. A dedicated service in Azure services portfolio by the name Azure AD Privileged Identity Management (PIM) manages JIT access, allowing administrators to control and monitor privileged access to Azure resources and Azure AD. PIM can generate alerts for suspicious or unsafe activities, enhancing security monitoring. This is commonly used for administrative tasks, accessing sensitive data, or managing critical infrastructure.

Just-in-time (JIT) access model is a security approach that grants privileged access or permissions only for the finite moments needed. It eliminates always-on, persistent privileged access, known as "standing privileges." On the other hand, Just Enough Access aka JEA model is essential for implementing the principle of least privilege. But "true least privilege" requires combining both models, so that organizations can minimize potential attackers' footholds and the paths to privilege that could escalate an attack. However, many enterprises struggle with having too many accounts with unnecessary privileges, standing access status quo, privilege blindness, and lack of context around privileged risk. By combining these approaches, organizations can significantly reduce the attack surface and minimize potential vulnerabilities. Some of the malpractices include deploying too many accounts with unnecessary privileges, permissions, and entitlements, a standing access status quo, privileged blindness, and lack of context around privileged risk.

In Amazon Web Services (AWS), limiting human access to cloud resources is crucial for security. AWS offers tools like AWS Identity and Access Management (IAM) and AWS IAM Identity Center for managing access. Granting just-in-time access to developers for a limited time based on approval is an effective way to limit active time frames for assignments to AWS resources. Okta's integration with IAM Identity Center allows customers to access AWS using their Okta identities. As an example, the roles could correspond to different job functions within your organization. For example, the "AWS EC2 Admin" role could correspond to a DevOps on-call site reliability engineer (SRE) lead, whereas the "AWS EC2 Read Only" role may apply to members of your development team. The step-by-step configuration for this involves setting up groups representing different privilege levels, enabling automatic provisioning of groups using SCIM protocol, assigning access for groups in Okta, creating permissions sets in IAM identity center, assign group access in your AWS organization, configuring Okta identity governance access requests and finally testing the configuration. Okta's integration with AWS minimizes persistent access assignments, granting access just in time for specific operational functions. This solution allows empty user groups to be assigned to highly-privileged AWS permissions, with Okta Access Requests controlling group membership duration. AWS supports Privileged Access Management aka PAM solutions where third-party solutions can be integrated into the AWS to provide ephemeral JIT access, ensuring that users only have the necessary privileges for the duration of their tasks. AWS provides regular fine-grained permissions for users, groups and roles with its Identity and Access Management policies which can even be used to restrict access to a certain time of the day. The single sign-on service can work with different identity providers to enforce JIT access. Finally, the AWS Security Token Service can issue temporary security credentials that provide limited time access to AWS resources.

In Azure, Conditional Access templates provide a convenient method to deploy new policies aligned with Microsoft recommendations. These templates are designed to provide maximum protection aligned with commonly used policies across various customer types and locations. The templates are organized into secure foundation, zero trust, remote work, protect administrator, and emerging threats. Certain accounts must be excluded from these templates such as emergency-access or break-glass accounts to prevent tenant-wide account lockout and some service accounts and service principals that are non-interactive and tied to any particular user.

To bolster the physical security, reducing the risk of malware or unauthorized access, streamlining and restricting activities that can be performed with the escalation of privilege, Microsoft hands out Secure Admin Workstations (SAWs) that are specialized and dedicated devices used exclusively for administrative tasks. They are particularly valuable in high-risk environments where security is paramount. Public clouds happen to be the most widely used cloud but there are other clouds that can be dedicated in scope specifically for governments, defense departments and those that require tighter access control and these are collectively called sovereign clouds. These clouds are especially benefited with SAW devices. Only authorized personnel can use SAWs, and they are often subject to strict security policies and monitoring. As an example, Microsoft uses approximately 35,000 SAW devices, with a small number dedicated to accessing these high-risk environments aka sovereign clouds.

Technical Debt in IaC

A case study might be a great introduction to this subject. A team in an enterprise wanted to set up a new network in compliance with the security standards of the organization and migrate resources from the existing network to the new one. When they started out allocating subnets from the virtual network address space and deploying the first few resources such as an analytical workspace and its dependencies, they found that the exact same method provisioning for the old network did not create a resource that was at par with the functionality of the old one. For example, a compute instance could not be provisioned into the workspace in the new subnet because there was an error message that said, "could not get workspace info, please check the virtual network and associated rules". It turned out that subnets were created with an old version of its definition from the IaC provider and lacked the new settings that were introduced more recently and were required for compatibility with the recent workspace definitions also published by the same IaC provider. The documentation on the IaC provider's website suggests that the public cloud that provides those resources had introduced breaking changes and newer versions required newer definitions. This forced the team to update the subnet definition in its IaC to the most recent from the provider and redo all the allocations and deployments after a tear down. Fortunately, the resources introduced to the new virtual network were only pilots and represented a tiny fraction of the bulk of the resources supporting the workloads to migrate.

The software engineering industry is rife with versioning problems in all artifacts that are published and maintained in a registry for public consumption ranging from diverse types as languages, packages, libraries, jars, vulnerability definitions, images, and such others. In the IaC, the challenge is somewhat different because deployments are usually tiered and the priority and severity of a technical debt differs from case to case with infrastructure teams maintaining a wide inventory of deployments, their constituent resources, and customers. It just so happens in this example that the failures are detected early, and the resolutions are narrow and specific, otherwise rehosting and much less restructuring are not easy tasks because they require complex deployments and steps.

While cost estimation, ROI and planning are as usual to any software engineering upgrades and project management, we have the advantage of breaking down deployments and their redeployments into contained boundaries so that they can be independently implemented and tested. Scoping and enumerating dependencies come with this way of handling the technical debt in IaC. A graph of dependencies between deployments can be immensely helpful to curate for efforts – both now and in the near future.

DevOps for IaC

As with any DevOps practice, the principles on which they are founded must always include a focus on people, process, and technology. With the help of Infrastructure-as-a-code and blueprints, resources, policies, and accesses can be packaged together and become a unit of provisioning the environment.

The DevOps Adoption RoadMap has evolved over time. What used to be Feature Driven Development around 1999 gave way to Lean thinking and Lean software development around 2003, which was followed by Product development flows in 2009 and Continuous Integration/Delivery in 2010. The DevOps Handbook and the DevOps Adoption Playbook are recent as of the last 5-6 years. Principles that inform practices that resolve challenges also align accordingly. For example, the elimination of risk happens with automated testing and deployments, and this resolves the manual testing, processes, deployments, and releases.

The people involved in bringing build and deployments to the cloud and making use of them instead of outdated and cumbersome enterprise systems must be given roles and clear separation of responsibility. For example, developers can initiate the promotion of code package to the next environment but only a set of people other than the developers must allow it to propagate to production systems and with signoffs. Fortunately, this is well-understood and there is existing software such as ITSM, ITBM, ITOM and CMDB. These are fancy acronyms for situations such as:

1. If you have a desired state you want to transition to, use a workflow,
2. If you have a problem, open a service ticket.
3. If you want orchestration and subscribe to events, use events monitoring and alerts.

4. If you want a logical model of the inventory, use a configuration management database.

Almost all IT businesses are concerned about ITOM such as with alerts and events, ITSM such as with incidents and service requests, and intelligence in operations. The only difference is that they have not been used or made available for our stated purposes, but this is still a good start.

The process that needs to be streamlined is unprecedented at this scale and sensitivity. The unnecessary control points, waste and overhead must be removed, and usability must be one of the foremost considerations for improving adoption.

The technology is inherently different between cloud and enterprise. While they have a lot in common when it comes to principles of storage, computing and networking, the division and organization in the cloud has many more knobs and levers that require due diligence.

These concerns around people, process and technology are what distinguishes and makes this landscape so fertile for improvements.

Top-Down vs Bottoms-up

Centralized planning has many benefits for infrastructure as evidenced by parallels in construction industry and public transportation. The top-down approach in this context typically refers to a method where policy decisions and strategies are formulated at a higher, often governmental, or organizational level, and then implemented down through various levels of the system. This approach contrasts with a bottom-up approach, where policies and strategies are developed based on input and feedback from lower levels, such as local communities or individual stakeholders.

Such an approach might involve:

Centralized Planning: High-level authorities set transportation policies and plans, which are then executed by regional or local agencies.

Regulation and Standards: Establishing uniform regulations and standards for transportation systems, which must be adhered to by all stakeholders.

Funding Allocation: Decisions on the allocation of funds for transportation projects are made at a higher level, often based on broader economic and policy goals.

This approach can ensure consistency and alignment with national or regional objectives, but it may also face challenges such as lack of local adaptability and slower response to specific local needs.

On the other hand, a bottom-up approach typically involves building and configuring resources starting from the lower levels of the infrastructure stack, often driven by the needs and inputs of individual teams or developers. This approach contrasts with a top-down approach, where decisions and designs are made at a higher organizational level and then implemented downwards.

This approach involves the following features:

Developer-Driven: Individual developers or teams have the autonomy to create and manage their own resources, such as virtual machines, databases, and networking components, based on their specific project requirements.

Incremental Development: Infrastructure is built incrementally, starting with basic components, and gradually adding more complex services and configurations as needed. This allows for flexibility and adaptability.

Agility and Innovation: Teams can experiment with new services and technologies without waiting for centralized approval, fostering innovation and rapid iteration.

Infrastructure as Code (IaC): Tools like Terraform and Azure Resource Manager (ARM) templates are often used to define and manage infrastructure programmatically. This allows for version control, repeatability, and collaboration.

Feedback Loops: Continuous feedback from the deployment and operation of resources helps teams to quickly identify and address issues, optimizing the infrastructure over time.

This approach can be particularly effective in dynamic environments where requirements change frequently, and rapid deployment and scaling are essential

The right approach depends on a blend of what suits the workloads demanded by the business and the system architecture that will best serve the organization in the long run across change in business requirements and directions.

Mitigation of political and regulatory risk in cloud infrastructure

Cloud infrastructure faces a shortfall in spending each year even with discounts from cloud providers and industry wide acceptance and corroboration of shift to cloud services. Within an organization, a closer inspection of this deficit often reveals political and regulatory hurdles that central versus individual teams' partnerships struggle to overcome. With a wide variety of cloud service portfolio constantly evolving and an even greater variety in their usages, an objective overview serves all stakeholders.

Application engineering and infrastructure consumers will often be reluctant to set the standards, maintain and enforce the infrastructure. Cloud resource management and cost optimization advisories from the cloud provider will continue to accrue and even necessitate immediate actions. The governance that an organization specifies in addition to the cloud service recommendations must minimize political distortions and allow independent investments by teams for their use cases. Guarantees and political-risk insurance such as pilot programs and sandbox environments can help overcome the challenges faced by early adopters. Company-wide agreements and standardization sought by security hardening initiatives can help with refining the policies enforced over the cloud infrastructure.

The lifespan of an infrastructure project moves through three distinct stages each with its own risks. First, there is the planning or design stage where a change of hands in the executive leadership might cancel projects that predecessors approved. Change in business needs, market pressure and sustainability drives might impose unanticipated conditions. Even the internal consumers within the company might ask for less than ideal alternatives. Second, change of ownership between department boundaries and organizational hierarchy can cause policy shifts that can dramatically affect costs. Lastly, as the infrastructures nears its end, there can be a lot of ambiguity and uncertainty on dates and mode of retirement including contention over its residual value. Decommissioning may even take a long time without efforts to reel it in.

There is no silver bullet to addressing many facets of such complex problems but documenting and continuously evaluating the state of the union and the forces at play will help with going fast or slow on certain key initiatives at the proper time. A series of examples, case studies and war stories often found in the cloud community can serve as examples to emulate which dispels any indecision and procrastination. It is a good thing if there can be proper representation of stakeholders and constant buy-ins on steps taken over time. This multilateral input together with an execution team nimble enough to deliver on objectives and key results aka OKRs will smoothen the experience with cloud infrastructure and return-on-investments. Finding a secure middle ground implies preparing projects rigorously, marketing investment opportunities, sounding out the market, and preparing comprehensive documentation.

Due to the huge amount of capital required for cloud resources, company designated oversight committees and enforcement team must prioritize infrastructure projects. When a detailed study is not possible, some "stopgap approach" is often called for. This approach might not provide a conclusive list of selections to work on, but it does offer valuable data and decision criteria. When there are multiple teams vying for the budget in the cloud, a new investment evaluation technique might be called for. This "Infrastructure Prioritization Framework" has many parallels in other industries and the central theme is to incorporate as much lateral feedback as possible to drive the best decisions. However, caution must be exercised against regressive biases creeping into the process such as tendency to favor projects in some business objectives merely because they score higher on some inputs. Evidence often suggests that infrastructure prioritization is based on politics, loose qualitative assessments, or professional judgement and without clear principles underpinning selection. On the other hand, a technique such as Analytic Hierarchy Process can help to organize and analyze complex decisions by introducing democracy in decision-making where every voice counts but some might count more based on expertise or relevance.

Feasibility studies and pilot programs can unearth facts beyond the marketing and documentation and can even provide a perspective into the net present value of different projects which is useful information to have to make good decisions. The IDF framework does rank projects for limited financing, but it incorporates the social, economic, environmental, and financial aspects into two indices: a socio-governmental index and a financial-economic index. This approach differs from other multi-criteria methods in that it weaves the various considerations and the developmental goals and policy objectives with a practical and cost-conscious approach where results of the analysis are often presented in charts and graphs. The format is flexible, and some judgement involved on what considerations to include but constant guarding against "regressive biases" is inevitable.

Data Engineering

The history of data engineering has evolved from big data frameworks like Hadoop and MapReduce to streamlined tools like Spark, Databricks, BigQuery, Redshift, Snowflake, Presto, Trino, and Athena. Cloud storage and transformation tools have made data more accessible, and lakehouses have offered a cost-efficient, unified option for managing data at scale. This evolution has led to a more accessible and efficient data management landscape.

Data transformation environments vary, with common environments being data warehouses, data lakes, and lakehouses. Data warehouses use SQL for transformation, while data lakes store large amounts of data economically. Lakehouses combine aspects of both, offering flexibility and cost-effectiveness. Databricks SQL is a serverless data warehouse that sits on the lakehouse platform. The choice between these environments depends on project needs, team expertise, and long-term data strategy.

Data staging is a crucial process in data transformation, often written in a temporary state to a suitable location, such as cloud storage or an intermediate table. Medallion architecture preserves data history and makes time travel possible. It comprises three distinct layers: Bronze for raw data, Silver for light transformation, and Gold for "clean" data. Bronze data is raw and unfiltered, Silver data is filtered, cleaned, and adjusted, and Gold data is stakeholder-ready and sometimes aggregated. This approach can be used in a lake or warehouse, breaking down each storage layer into discrete stages of data cleanliness.

Data transformation is largely influenced by the tools available, with Python being a popular choice in the digital era. Python's Pandas library is at its core, and it has evolved significantly in data processing. However, scaling Python for large datasets has been challenging, often requiring libraries like Dask and Ray. Python-based data processing is a renaissance, with Rust emerging. To transform data in Python, choose a suitable library and framework, such as Pandas or emerging libraries like Polars and DuckDB. SQL, a declarative language, can be used as a declarative or imperative language, but is limited by a lack of functionality. Languages like Jinja/Python and JavaScript often complement SQL workflows. Rust, a new transformation language, is considered the future of data engineering, but Python has a solid foothold due to its community support and library ecosystem.

Transformation frameworks are multilanguage engines for executing data transformations across machines or clusters, enabling transformations to be manipulated in various languages like Python or SQL. Two popular engines are Hadoop and Spark. Hadoop, an open-source framework, gained traction in the mid-2000s with tech giants like Yahoo, Facebook, and Google. However, its MapReduce was not well-suited for real-time or iterative workloads, leading to the rise of Apache Spark in the early 2010s. Spark, a powerful open-source data processing framework, revolutionized big data analytics by offering speed, versatility, and integration with key technologies. Its key innovation is resilient distributed datasets (RDDs), enabling in-memory data processing and faster computations. With the rise of serverless data warehouses, big data engines may no longer be necessary, but query engines like BigQuery, Databricks SQL, and Redshift should not be disregarded. Recent advancements in in-memory computation may continue to expand data warehouses' transformation capabilities.

Data transformation is a crucial process that involves pattern mapping and understanding the different transformations that should be applied. Enrichment involves enhancing existing data with additional sources, such as adding demographic information to customer records. Joining involves combining two or more datasets based on a common field, like a JOIN operation in SQL. Filtering selects only the necessary data points for analysis based on certain criteria, reducing the volume, and improving the quality of the data. Structuring involves translating data into a required format or structure, such as transforming JSON documents into tabular format or vice versa. Conversion is changing the data type of a particular column or field, especially when converting between semi-structured and structured data sources. Aggregation is summarizing and combining data to draw conclusions from large volumes of data, enabling insights to inform business decisions and create value from data assets. Anonymization is masking or obfuscating sensitive information within a dataset to protect privacy. It involves hashing emails or removing personally identifiable information (PII) from records. Splitting is a form of denormalization, dividing a complex data column into multiple columns. Deduplication is the process of removing redundant records to create a unique dataset, often through aggregation, filtering, or other methods.

Data update patterns are essential for transforming data in a target system. Overwrite is the simplest form, which involves a complete drop of an existing source or table and an overwrite with new data. Inserting is a more complex pattern, involving the appending of new data to an existing source without changing existing rows. Upsert is a more complex pattern, with applications for change data capture, sessionization, and deduplication. Platforms like Databricks have MERGE functionality to simplify the process. Data deletion is often misunderstood, with two main types: "hard" and "soft." Soft deletes enable the creation of historical records for an asset's status, while hard deletes eliminate these records, which can be problematic in data recovery cases.

When building a data transformation solution, consider several best practices, including staging, idempotency, normalization, and incrementality. Staging protects against data loss and ensures a low time to recovery (TTR) in case of failure. Idempotency ensures consistency and reliability by performing something multiple times, similar to reproducibility. Normalization refines data to a clean, orderly format, while denormalization duplicates records and information for improved performance. Incrementality determines whether a pipeline is a simple INSERT OVERWRITE or a more complex UPSERT. Predefined patterns for building incremental workflows can be found in tools like dbt and Airflow. Real-time data transformation involves batch, micro-batch, and streaming transformations. Micro-batch approaches, like Apache Spark's PySpark and Spark SQL, are simpler to implement compared to true, single-event transformations. Spark Structured Streaming is a popular streaming application that efficiently handles incremental and continuous updates, achieving latencies as low as 100 milliseconds with exactly once fault tolerance. Continuous Processing, introduced in Spark 2.3, can reduce latencies to as little as 1 millisecond, further enhancing its capability for streaming data transformation.

The modern data stack is experiencing a second renaissance due to new technologies and AI advancements. As a result, new tools and technologies are emerging to redefine data transformation. However, it's crucial to adhere to timeless strategies for managing data and creating cleaned assets. Supercharged tooling and automations can be both beneficial and challenging, but engineers must ensure well-planned and executed transformation systems with a high value-to-cost ratio.

Dependency management and pipeline orchestration are aptly dubbed "under-currents" in the book "Fundamentals of Data Engineering" by Matt Housley and Joe Reis.

Data orchestration is a process of dependency management, facilitated through automation. It involves scheduling, triggering, monitoring, and resource allocation. Data orchestrators are different from schedulers, which are cron-based. They can trigger events, webhooks, schedules, and even intra-workflow dependencies. Data orchestration provides a structured, automated, and efficient way to handle large-scale data from diverse sources.

Orchestration steers workflows toward efficiency and functionality, with an orchestrator serving as the tool enabling these workflows. They typically trigger pipelines based on a schedule or a specific event. Event-driven pipelines are beneficial for handling unpredictable data or resource-intensive jobs.

The perks of having an orchestrator in your data engineering toolkit include workflow management, automation, error handling and recovery, monitoring and alerting, and resource optimization. Directed Acyclic Graphs or DAGs for short bring order, control, and repeatability to data workflows, managing dependencies and ensuring a structured and predictable flow of data. They are pivotal for orchestrating and visualizing pipelines, making them indispensable in managing complex workflows, particularly within a team or large-scale setups. For example, a DAG serves as a clear roadmap defining the order of tasks and with this lens, it is possible to organize the creation, scheduling, and monitoring of data pipelines.

Data orchestration tools have evolved significantly over the past few decades, with the creation of Apache Airflow and Luigi being the most dominant tools. However, it is crucial to choose the right tool for the job, as each tool has its strengths and weaknesses. Data orchestrators, like the conductor of an orchestra, balance declarative and imperative frameworks to provide flexibility and efficiency in software engineering best practices.

When selecting an orchestrator, factors such as scalability, code and configuration reusability, and the ability to handle complex logic and dependencies are important to consider. The orchestrator should be able to scale vertically or horizontally, ensuring that the process of orchestrating data is separate from the process of transforming data.

Orchestration is about connections, and platforms like Azure, Amazon, and Google offer hosted versions of popular tools like Airflow. Platform-embedded alternatives like Databricks Workflows provide more visibility into tasks orchestrated by orchestrators. Popular orchestrators have strong community support and continuous development, ensuring they remain up to date with the latest technologies and best practices. Support is crucial for both closed source and paid solutions, as solutions engineers can help resolve issues.

Observability is essential for understanding transformation flows and ensuring your orchestrator supports various methods for alerting your team. To implement an orchestration solution, you can build a solution, buy an off-the-shelf tool, self-host an open-source tool, or use a tool included with your cloud provider or data platform. Apache Airflow, developed by Airbnb, is a popular choice due to its ease of adoption, simple deployment, and ubiquity in the data space. However, it has flaws, such as being engineered to orchestrate, not transform or ingest.

Open-source tools like Airflow and Prefect are popular orchestrators with paid, hosted services and support. Newer tools like Mage, Keboola, and Kestra are also innovating. Open-source tools offer community support and the ability to modify source code. However, they depend on support for continued development and may risk project abandonment or instability. A tool's history, support, and stability must be considered when choosing a solution.

Data orchestration is a crucial aspect of modern data engineering, involving the use of relational databases for data transformation. Tools like dbt, Delta Live Tables, Dataform, and SQLMesh are used as orchestrators to evaluate dependencies, optimize, and execute commands against a database to produce desired results. However, there is a potential limitation in data orchestration due to the need for a mechanism to observe data across different layers, leading to a disconnection between sources and cleaned data. This can be a challenge in identifying errors in downstream data.

Design patterns can significantly enhance the efficiency, reliability, and maintainability of data orchestration processes. Some orchestration solutions make these patterns easier, such as building backfill logic into pipelines, ensuring idempotence, and event-driven data orchestration. These patterns can help avoid one-time thinking and ensure consistent results in data engineering. Choosing a platform-specific data orchestrator can provide greater visibility between and within data workflows, making it essential for ETL workflows.

Orchestrators are complex and difficult to develop locally due to their complex trigger actions. To improve performance, invest in tools that allow for fast feedback loops, error identification, and a local environment that is developer friendly. Retry and fallback logic are essential for handling failures in a complex data stack, ensuring data integrity and system reliability. Idempotent pipelines set up scenarios for retrying operations or skipping and alerting the proper parties. Parameterized execution allows for more malleability in orchestration, allowing for multiple cases and reuse of pipelines. Lineage refers to the path traveled by data through its lifecycle, and a robust lineage solution is crucial for debugging issues and extending pipelines. Column-level lineage is becoming an industry norm in SQL orchestration, and platform-integrated orchestration solutions like Databricks Unity Catalog and Delta Live Tables offer advanced lineage capabilities. Pipeline decomposition breaks pipelines into smaller tasks for better monitoring, error handling, and scalability. Building autonomous DAGs can mitigate dependencies and critical failures, making it easier to build and debug workflows.

The evolution of transformation tools, such as containerized infrastructure and hybrid tools like Prefect and Dagster, may change the landscape of data teams. These tools can save time and resources, enabling better observability and monitoring within data warehouses. Emerging tools like SQLMesh may challenge established players like dbt, while plug-and-play solutions like Databricks Workflows are becoming more appealing. These developments will enable data teams to deliver quality data in a timely and robust manner.

The return on investment in data engineering projects is often reduced by how fragile the system becomes and the maintenance it requires. Systems do fail but planning for failure means making it easier to maintain and extend, providing automation for handling errors and learning from experience. The minimum viable product principle and the 80/20 principle are time-honored traditions.

Direct and indirect costs of ETL systems are significant, as they can lead to inefficient operations, long run times, and high bills from providers. Indirect costs, such as constant triaging and data failure, can be even more significant. Teams that win build efficient systems that allow them to focus on feature development and data democratization. SaaS (Software as a Service) offers a cost-effective solution, but it can also lead to loss of trust, revenue, and reputation. To minimize these costs, focus on maintainability, data quality, error handling, and improved workflows. Monitoring and benchmarking are essential for minimizing pipeline issues and expediting troubleshooting efforts. Proper monitoring and alerting can help improve maintainability of data systems and lower costs associated with broken data. Observing data across ingestion, transformation, and storage, handling errors as they arise, and alerting the team when things break, is crucial for ensuring good business decisions.

Data reliability and usefulness are assessed using metrics such as freshness, volume, and quality. Freshness measures the timeliness and relevance of data, ensuring accurate and recent information for analytics, decision-making, and other data-driven processes. Common metrics include the length between the most recent timestamp and the current timestamp, lag between source data and the dataset, refresh rate, and latency. Volume refers to the amount of data needed for processing, storage, and management within a system. Quality involves ensuring data is accurate, consistent, and reliable throughout its lifecycle. Examples of data quality metrics include uniqueness, completeness, and validity.

Monitoring involves detecting errors in a timely fashion and implementing strict measures to improve data quality. Techniques to improve data quality include logging and monitoring, lineage, and visual representations of pipelines and systems. Lineage should be complete and granular, allowing for better insight and efficiency in triaging errors and improving productivity. Overall, implementing these metrics helps ensure data quality and reliability within an organization.

Anomaly detection systems analyze time series data to make statistical forecasts within a certain confidence interval. They can catch errors that might originate outside the systems, such as a bug in a payments processing team that decreases purchases. Data diffs are systems that report on data changes presented by changes in code, ensuring accurate systems remain accurate especially when used as an indicator on data quality. Tools like Datafold and SQLMesh have data diffing functionality. Assertions are constraints put on data outputs to validate source data. They are simpler than anomaly detection and can be found in libraries like Great Expectations aka GX Expectations or systems with built-in assertion definitions.

Error handling is crucial for data systems and recovery from their effects, such as lost data or downtime. Error handling involves automating error responses or boundary conditions to keep data systems functioning or alert the team in a timely and discreet manner. Approaches include conditional logic, retry mechanisms, and pipeline decomposition. These methods help keep the impact of errors contained and ensure the smooth functioning of data systems.

Graceful degradation and error isolation are essential for maintaining limited functionality even when a part of a system fails. Error isolation is enabled through pipeline decomposition, which allows systems to fail in a contained manner. Graceful degradation maintains limited functionality even when a part of the system fails, allowing only one part of the business to notice an error.

Alerting should be a last line of defense, as receiving alerts is reactive. Isolating errors and building systems that degrade gracefully can reduce alarm fatigue and create a good developer experience for the team.

Recovery systems should be built for disasters, including lost data. Staged data, such as Parquet-based formats like Delta Lake and patterns like the medallion architecture, can be used for disaster recovery. Backfilling, the practice of simulating historical runs of a pipeline to create a complete dataset, can save time when something breaks.

Improving workflows is crucial in data engineering, as it is an inherently collaborative job. Data engineering is a question of when things break, not if. Starting with systems that prioritize troubleshooting, adaptability, and recovery can reduce headaches down the line.

In the context of software teams, understanding their motivations and workflows is crucial for fostering healthy relationships and improving efficiency. By focusing on the team's goals and understanding their workflows, you can craft a process to improve efficiency.

Structured, pragmatic approaches can ensure healthy relationships through Service-Level Agreements (SLAs), data contracts, APIs, compassion, and empathy, and aligning incentives. SLAs can be used to define performance metrics, responsibilities, response and resolution times, and escalation procedures, improving the quality of data that is outside of your control. Data contracts, popularized by dbt, govern data ingested from external sources, providing a layer of standardization and consistency. APIs can be used to transmit an expected set of data, providing granular access control, scalability benefits, and versioning, which can be useful for compliance.

Compassion and empathy are essential in engineering and psychology, as understanding coworkers' motivations, pain points, and workflows allows for effective communication and appeal to their incentives. In the digital age, it's essential to go the extra mile to understand coworkers and appeal to their incentives.

Setting key performance indicators (KPIs) around common incident management metrics can help justify the time and energy required to do the job right. These metrics include the number of incidents, time to detection (TTD), time to resolution (TTR), and data downtime (N × [TTD + TTR].

Continually iterating and adjusting processes in the wake of failures and enhancing good pipelines to become great are some ways to improve outcomes. Documentation is crucial for understanding how to fix errors and improve the quality of data pipelines. Postmortems are valuable for analyzing failures and learning from them, leading to fewer events that require recovery. Unit tests are essential for validating small pieces of code and ensuring they produce desired results. Continuous integration/continuous deployment (CI/CD) is a preventative practice to minimize future errors and ensure a consistent code base.

Engineers should simplify and abstract complex code to improve collaboration and reduce errors. Building data systems as code, which can be rolled back and reverted to previous states, can improve observability, disaster recovery, and collaboration. Tools that are difficult or impossible to version control or manipulate through code should be exercised with caution. With responsibilities defined, incentives aligned, and a monitoring/troubleshooting toolkit, engineers can automate and optimize data workflows. Balancing automation and practicality is essential in data engineering, ensuring robust, resilient systems ready for scaling.

Support

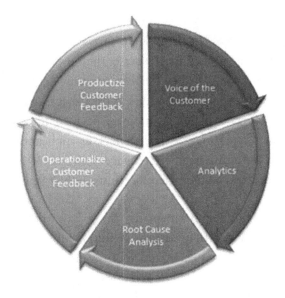

As with all software, infrastructure-as-code must strive to meet the expectations of the customers. Deployments and configurations are not always fully suited to a customer at the time of hand-off and workloads themselves undergo changes during the lifetime of the deployments. Infrastructure engineers often compensate for this with issue tracking and maintaining multiple instances of deployment usually differing only in their suffix as called out in the naming convention. A higher suffix indicates a newer version of the same deployment that was commissioned for a specific purpose. The older versions are eventually decommissioned if they cannot be properly tuned to close the gap between the customers' expectations and the current configuration. If all else fails and the support from company internal resources and external support from the public cloud fails, the infrastructure customers are eased into redeployments.

When infrastructure engineers are faced with support calls for their customers, there are several factors that play out into the extent of maintenance engaged such as revenue, customer segmentation, market segment, costs, resources, media, etc. and it is not uncommon to find a grading in services offered. Support is all about this art and science of engaging with customers throughout their usages. It is interesting to note that customers can run into issues of their own accord with any of these versions and not just when the infrastructure engineer has put out a deployment that the customer wants to use. That is why management and operations are both ongoing commitments for a infrastructure while being fundamentally different. In business, as application engineering team and infrastructure engineers engage, there is often a funny and cheeky practice that could be mentioned in a lighter vein.

Deployments that suffer from quality often have more maintenance that incurs a lot of costs in terms of efforts, time and money. One way to address this is to pass it on to customers! Customers may be ready to pay a price but often demand a price from the deployers for not putting the defects in the first place! This conversation could have different connotations if the size of the giver or taker were skewed. For example, this could imply negotiations or coercions, pricing, penalties and differentiation.

As engineers remain open to understanding the needs on a case-by-case basis, some best practices in engineering, operations and management continue to prevail for the greater good. As such cases become an interesting insight into infrastructure itself, it provides a learning that can be incorporated to improve the system and not to try to beat it.

Printed in the United States
by Baker & Taylor Publisher Services